Stochastic integration and generalized martingales

A U Kussmaul

University of Hull

Stochastic integration and generalized martingales

Pitman Publishing
LONDON · SAN FRANCISCO · MELBOURNE

PITMAN PUBLISHING LIMITED
39 Parker Street, London WC2B 5PB

PITMAN PUBLISHING CORPORATION
6 Davis Drive, Belmont, California 94002, USA

Associated Companies
Copp Clark Ltd, Toronto · Fearon Publishers Inc, Belmont, California
Pitman Publishing Co. SA (Pty) Ltd, Johannesburg · Pitman Publishing
New Zealand Ltd, Wellington · Pitman Publishing Pty Ltd, Melbourne
Sir Isaac Pitman Ltd, Nairobi

AMS Subject Classifications: (main) 60–02, 60H05
(subsidiary) 60–01, 60G45, 60G40,
46–01, 46G10

Library of Congress Cataloging in Publication Data

Kussmaul, A U 1947–
  Stochastic integration and generalized martingales.

  (Research notes in mathematics; 11)
  Includes bibliographical references and index.
  1. Integrals, Stochastic. 2. Martingales (Mathematics)
I. Title. II. Series.
QA274.22.K87   519.2   76–30277
ISBN 0–273–01030–1

© A U Kussmaul 1977

All rights reserved. No part of this publication may be reproduced,
stored in a retrieval system, or transmitted, in any form or by any
means, electronic, mechanical, photocopying, recording and/or
otherwise without the prior written permission of the publishers.

Reproduced and printed by photolithography
in Great Britain at Biddles of Guildford

Preface

In recent years the mathematical theory of stochastic integration has become of interest because of its use in several areas of mathematical physics.

Stochastic integrals with respect to martingales were first discussed by Wiener [27], who defined the integral for deterministic integrands. The extension of this definition to integrands which are stochastic processes is due to Ito in the Brownian motion case (see [13]) and to Doob ([6]) for the case of square integrable martingales. Subsequently the theory and the calculus of stochastic integration have been developed considerably, mainly by the French school of probabilits (see [15], [20], [23]).

There are basically two ways of defining the stochastic integral with respect to martingales, the first one exploits generalizations of the quadratic variation of square integrable martingales (see [20]), the second one uses vector valued measures (see [23]). In the present monograph the latter approach is developed to the extent that it applies to quasimartingales which are dominated in the modulus by an integrable function. At the same time it is shown that the first approach is dual to the second one in the sense that a measure M on a σ-algebra Σ with values in a Banach space E can equivalently be described by the map M', which maps elements x' of the dual E' of E to the real measures $M'(x')(A) = \langle M(A), x' \rangle$ $(A \in \Sigma)$. The main incentive for writing this account was to point out the relation between these approaches and to imbed the theory of stochastic integration into a functional analytic framework.

In this book we develop the theory of the stochastic integral together with the necessary prerequisites from the theory of stochastic processes in one volume, thus forming an introduction to this subject for readers familiar with Banach spaces and with measure theory.

The book contains three chapters, divided into sections which are numbered consecutively. Chapter 1 introduces the notation and basic properties of stochastic processes; section 2, dealing with integration with respect to processes of bounded variation, is intended to serve as introduction to the theory of the stochastic integral. Chapter 2 is concerned with general theory of processes, and with martingales and quasimartingales, processes which form the main tool for the theory of the stochastic integral. In sections 4 and 5 classical results of the theory of martingales, such as regularity properties of paths, optional stopping and martingale inequalities are presented. Section 4 also contains more recent results about the space BMO of martingales of bounded mean oscillation and its predual. In section 6 stopping times are investigated, in particular stopping times which can be approximated from below by strictly smaller stopping times. In section 7 the σ-algebra of predictable sets is introduced and studied. Stochastic processes, measurable with respect to this σ-algebra, serve as integrands for the stochastic integral. The projection theorem of section 7 is used in section 8 to investigate the correspondence between processes of integrable variation and the real measures generated by them in much greater detail than was done in section 2. Finally in section 9 the structure of quasimartingales is studied.

The last chapter starts with a preliminary section about vector valued measures, establishing extension theorems and some aspects of the integration theory, needed in the sequel. In section 11 p-summable processes are introduced, i.e. stochastic processes giving rise in a natural way to a σ-additive measure, defined on the σ-algebra of predictable sets, with values in the Banach spaces $L^p(P)$ $(1 \leq p \leq \infty)$. This is the class of processes with respect to which the stochastic integral is defined. Square integrable martingales are studied as examples of 2-summable processes. The concept of localization, introduced in section 12, allows us to define stochastic integrals with respect to semimartingales. In the second part of this section the Banach space SP of summable (i.e. 1-summable) stochastic processes is studied and summable processes are identified as quasimartingales which are dominated in modulus by an integrable function. The latter result is obtained by a new characterization of the space H^1 of martingales as the space of summable martingales, thus pointing out the importance of this space for the theory of the stochastic integral (compare [20] chap. V). The final section is concerned with the space SM of summable martingales and with applications of the theory to questions related to the representation of martingales as stochastic integrals.

It should be pointed out that this book does not aspire to be a complete account of stochastic integration. In particular the calculus of the stochastic integral, based on the famous Ito formula, with its applications in the theory of stochastic differential equations is not treated at all. Our objective has been to illucidate the somewhat neglected foundations of stochastic integration, and to stimulate further research in this area.

I would like to thank Professor R.J. Elliott for his encouragement to write this book and Professor S.J. Taylor for many helpful suggestions for improving the manuscript. Finally I acknowledge with thanks that, while writing these notes, I was supported by a grant from the Science Research Council.

Alfred Kussmaul
University of Hull,
November 1975.

Contents

1 Preliminaries

 §1 Notation and basic definitions 1

 §2 Integration with respect to stochastic processes of bounded variation 5

 §3 Adapted processes and stopping times 11

2 Generalised martingales and general theory of processes

 §4 Discrete martingales 16

 §5 Continuous parameter martingales 39

 §6 Classification of stopping times 44

 §7 Basic subσ-algebras of $\Sigma = B \otimes F$ and projections of stochastic processes 52

 §8 Processes of bounded variation and projections 69

 §9 Quasimartingales 80

3 Stochastic integration

 §10 Integration with respect to measures with values in a Banach space 99

 §11 Stochastic integration with respect to square integrable martingales and martingales of bounded variation 112

 §12 Locally summable and summable stochastic processes 124

 §13 Decomposition of the space SM of summable martingales and representation of martingales as stochastic integrals 142

References 155

Index of symbols 158

Subject index 161

1 Preliminaries

§1 NOTATION AND BASIC DEFINITIONS

1.1 Measure Spaces: Let Ω be a set, Σ a ring of subsets of Ω.

An additive set function on Σ with values in a real vectorspace E is called *E-valued measure* (or simply measure if there is no danger of confusion) on Σ.

If E is the vector space R of real numbers, we write real measure instead of R-valued measure and positive measure if the set function has its values in the set of positive real numbers R_+.

A positive σ-additive measure with total mass 1, defined on a σ-algebra Σ, is called probability measure and denoted by P throughout the monograph.

Real measures are denoted by greek letters μ, ψ, λ; vector space valued measures usually by capital letters.

The triple (Ω, Σ, P), where P is a probability measure is called *probability space*, we denote by $L^p(P)$ ($1 \leq p \leq \infty$) the spaces of p-integrable (resp. essentially bounded) functions, or more precisely of equivalence classes of such functions.

Elements of $L^p(P)$ are usually denoted by small letters f, g, h.

A probability space is called *complete*, if Σ contains all P-null sets.

For an element $f \in L^p(P)$, we denote $\int f \, dP$ by $E(f)$ and we write $E(f|\Sigma_o)$ for the conditional expectation of f with respect to a sub σ-algebra $\Sigma_o \subseteq \Sigma$. For $1 \leq p < \infty$ the norm $f \to E(|f|^p)^{1/p}$ on $L^p(P)$ is denoted by $\|f\|_p$, the essential supremum norm on $L^\infty(P)$ by $\|f\|_\infty$.

1.2 Stochastic Processes:

Let (Ω, F, P) be a probability space, S a set and A a σ-algebra of subsets of S.

A mapping $X: T \times \Omega \to S$, where T is a subset of the extended positive real line $\overline{R}_+$, is called a *stochastic process* if for every $t \in T$ the map $\omega \to X(t, \omega)$ is an S-valued random variable.

(Ω, F, P) is called the *base* of the process X, (S, A) the state space of X.

We denote the map $\omega \to X(t, \omega)$ by X_t and write $X = (X_t)_{t \in T}$ or simply $X = (X_t)$.

For fixed $\omega \in \Omega$ the map $t \to X(t, \omega)$ is called the *path* $X(\omega)$ of the process $X = (X_t)$.

If S is a topological space, then X is called a *continuous process*, if P-almost surely all paths are continuous for the induced topology on T. Similarly we define *right continuous* and *left continuous* processes and the property that a process has *left hand* (resp. *right hand*) *limits*. If almost all paths are constant functions, X is called a *constant process*.

A stochastic process X is called *measurable* if the map X is measurable with respect to the σ-algebra $B \otimes F$, where B is the Borel σ-algebra on T and $B \otimes F$ is the product σ-algebra of B and F.

Two processes X and Y are called *indistinguishable* if P-almost surely all paths of X and Y coincide.

In the sequel we shall consider equivalence classes of indistinguishable processes rather than individual processes, but by an abuse of language we shall call these objects stochastic processes.

In this context it is clear that a stochastic process X is well defined if for P-almost all $\omega \in \Omega$ the paths $X(\omega)$ are defined.

A stochastic process $X = (X_t)$, such that P-almost all paths are real functions is called a *real process*. A real stochastic process $X = (X_t)$ will be called *evanescent* if X is indistinguishable from the process which is identically equal to zero, and a *set* $A \subseteq T \times \Omega$ is *evanescent* if the process X_A is evanscent.

The family of (equivalence classes of) real processes forms a vector space. In the sequel we shall only deal with real processes, the word 'real' will be omitted in future.

1.3 *E-processes*: Every stochastic process can be considered as a map $X: T \to M(F)$ from T into the vector space $M(F)$ of real random variables on (Ω, F), or more precisely, equivalence classes of real random variables with respect to the equality P-almost surely.

We call a map $X: T \to E$, where E is a real vector space, a *E-process*.

If E is a space of equivalence classes of random variables (with respect to the equivalence relation $Y \sim Z$ if and only if $Y = Z$ P-almost surely) and if we pick for each $t \in T$ a representative X_t of $X(t)$, the stochastic process $(X_t)_{t \in T}$ will be called a *modification* of the E-process X.

If on the other hand a given stochastic process $X = (X_t)$ is such that for every $t \in T$ X_t is a representative of some element of E, and if we want to stress the fact that X is considered as a map from T to E, we call X an E-process. In particular in order to distinguish between continuity properties of paths and continuity properties as an E-process, we say that $X = (X_t)$ is E-continuous, if X is considered as E-process, and E is a topological space. Throughout the monograph E will be one of the Banach spaces $L^p(P)$ ($1 \leq p \leq \infty$) or the space $M(F)$, equipped with the topology of convergence in P-measure.

The mapping which associates to every stochastic process $X = (X_t)$ the $M(F)$-process $t \to X_t$ is a linear map from the vectorspace of stochastic processes to the vector space of $M(F)$-processes. Keeping in mind that indistinguishable stochastic process are indentified, the obvious fact that two right (resp. left) continuous modifications of the same $M(F)$-process are indistinguishable, allows us to define a linear retract from the subspace of all $M(F)$-processes, which admit a right (resp. left) continuous modification to the vector space of all right (resp. left) continuous stochastic processes.

This is one of the reasons for the right continuity assumption throughout the monograph.

§2 INTEGRATION WITH RESPECT TO STOCHASTIC PROCESSES OF BOUNDED VARIATION

In this section the simplest case of stochastic integration will be studied. Although the definition of the stochastic integral with respect to a process of bounded variation will be straightforward, the various aspects of this integral will be of basic importance for the definition of a stochastic integral for a much larger class of stochastic processes.

2.2 Definition: A stochastic process $V = (V_t)_{t \in R_+}$ is called a process of *bounded variation*, if V satisfies the following assertions:

(a) V is measurable and right continuous.

(b) For every bounded interval $[o,t] \subseteq R_+$ the paths of V are P-almost surely of bounded variation on $[o,t]$.

A process of bounded variation is called *increasing*, if P-almost surely all paths of V are positive increasing functions.

Given a bounded measurable process $Z = (Z_t)$, we define the process $V(Z) = (V(Z)_t)$ by:

$$V(Z)(t,\omega) = \int_{[o,t]} Z_s(\omega)\, dV_s(\omega),$$

where the integral is the Stieltjes integral for every $\omega \in \Omega$. The process $V(Z) = (V(Z)_t)$ will be called the *stochastic integral* of Z with respect to V. As a matter of convenience, we shall occasionally write

$$\int_{[o,t]} Z_s dV_s \quad \text{instead of} \quad V(Z)_t.$$

If $V = (V_t)$ is a process of bounded variation, we denote by $|V|_t(\omega)$ the variation of the path $V_t(\omega)$ on the interval $[o,t]$, and write $|V| = (|V|_t)$ for the corresponding process, which is an increasing process.

If we implement the following restriction on the processes of bounded variation, we shall see that the stochastic integral can be interpreted as an integral with respect to a σ-additive real measure:

2.3 Definition: A process $V = (V_t)$ of bounded variation will be called *of integrable variation* if $E(|V|_\infty) < \infty$ holds, where $|V|_\infty(\omega)$ is the total variation of the path $V_t(\omega)$.

The vector space of processes of integrable variation is denoted by SV. It is easy to see that the functional $V \to E(|V|_\infty)$ defines a norm on SV, under which SV is a Banach space.

If we define for a measurable process $Z = (Z_t)$ the stochastic integral $V(Z) = (V(Z)_t)$ as before, the process $V(Z)$ is of integrable variation if and only if Z satisfies the condition $E(\int |Z_t| d|V|_t) < \infty$. Measurable processes Z, satisfying this condition, will be called *integrable (with respect to V)*.

Let now A be an element of the σ-algebra $\Sigma = B \otimes F$ and $V \in SV$ a process of integrable variation. Clearly the measurable process χ_A is integrable with respect to V, and by virtue of Fubini's theorem, the set function μ_V on Σ, defined as $\mu_V(A) = E(V(\chi_A)_\infty)$ for $A \in \Sigma$, is a σ-additive real measure on Σ.

(For a process of integrable variation $V \in SV$ and a V-integrable process Z the integrable random variable $V(Z)_\infty$ is defined as

$$V(Z)_\infty(\omega) = \int Z_s(\omega) dV_s(\omega)).$$

We call the measure μ_V the *real measure, generated by* $V \in SV$. If Σ' is a sub σ-algebra of Σ, the restriction of μ_V to Σ' will be called the *real measure generated by* V *on* Σ'.

It is obvious that the total variation $|\mu_V|(R_+ \times \Omega)$ of μ_V is equal to the norm $E(|V|_\infty)$ of V, and that every evanescent subset of $R_+ \times \Omega$ is a

μ_V-null set.

As we shall see later, the latter property characterizes measures occurring in stochastic integration, we therefore define:

2.4 Definition: Let Σ_0 be a sub σ-algebra of Σ. Every σ-additive (real or vectorvalued) measure μ on Σ_0, such that the evanescent subsets of $R_+ \times \Omega$ are μ-null sets, is called a *stochastic measure*.

For the following proposition keep in mind that uniqueness in connection with stochastic processes always means uniqueness up to indistinguishability.

We write $sca(\Sigma)$ for the Banachspace (under the total variation norm) of real stochastic measures on Σ. Notice that $sca(\Sigma)$ is a Banach lattice under the usual ordering.

2.5 Proposition: Let μ be a stochastic measure on the σ-algebra Σ of subsets of $R_+ \times \Omega$. There exists a unique stochastic process $V = (V_t) \in SV$ of bounded variation, such that μ and the measure μ_V, generated by V, coincide.

Proof: Since the positive part μ^+ of μ and the negative part μ^- of μ are elements of $sca(\Sigma)$, we may assume that μ is positive.

For every $t \in R_+$ define a measure μ^t on F by $\mu^t(F) = \mu([0,t] \times F)$ ($F \in F$). Since μ is σ-additive and doesn't charge evanescent subsets of $R_+ \times \Omega$, μ^t is a P-absolutely continuous measure for every $t \in R_+$, and we define the $L^1(P)$-process $\hat{V} = (\hat{V}_t)$ by $\hat{V}_t = d\mu^t/dP$, where $d\mu^t/dP$ is the Radon-Nikodym derivative of μ^t with respect to P.

In view of the positivity and the σ-additivity of μ, we have $\hat{V}_t \leq \hat{V}_s$ (in $L^1(P)$) for $t \leq s$ and $\hat{V} = (\hat{V}_t)$ is a $L^1(P)$-continuous process.

For every positive rational number $s \in Q$ let V_s be a representative of $\hat{V}_s$, and put $V_t(\omega) = \lim\limits_{s \downarrow t, s \in Q} V_s(\omega)$ for $t \in R_+$.

This limit exists P-almost everywhere and from the preceding observations we conclude that $V = (V_t)$ is a right continuous modification of the $L^1(P)$-process $\hat{V}$.

For every element $F \in \mathcal{F}$ we have
$$\mu_V([o,t] \times F) = E(\chi_F V_t) = \mu([o,t] \times F),$$
which implies that $\mu_V = \mu$ holds.

Finally the uniqueness is obvious from the above argument.

The Banach space SV of processes of integrable variation is an ordered vector space under the order relation $V \geq o$ if and only if $V \in SV$ is an increasing process. It is an immediate consequence from the definition of μ_V that V is increasing if and only if μ_V is positive. Together with Proposition 2.5 this yields the following theorem:

<u>2.6 Theorem</u>: The Banach spaces SV and $sca(\Sigma)$ are isometric isomorphic under the mapping which maps the stochastic process $V = (V_t) \in SV$ to the measure $\mu_V \in sca(\Sigma)$, generated by V.

The Banach space SV ordered by the positive cone $\{V \in SV : V$ is an increasing process$\}$ is a Banach lattice and the isomorphism defined above is a Banach lattice isomorphism between SV and $sca(\Sigma)$.

The 1-1 correspondence between SV and $sca(\Sigma)$, given by Theorem 2.6 allows a completely equivalent description of stochastic integration, as defined in 2.3, in terms of measures in $sca(\Sigma)$.

For a measurable process $Z = (Z_t)$ the assertion 'integrable with respect to V' corresponds to 'Z is μ_V-integrable', and the process $V(Z) \in SV$ is the process which generates the measure $\nu = Z\mu \in sca(\Sigma)$.

To conclude this section, we shall study another aspect of the stochastic integral, defined in 2.3. Let us first consider an increasing process of integrable variation $V \in SV$. The stochastic integral defines an $L^1(P)$-valued measure on Σ by $A \to V(\chi_A)_\infty$ for elements $A \in \Sigma$. We denote this measure by I_V. Because the $L^1(P)$-norm is additive on the positive cone of $L^1(P)$, the real measure μ_V is the variation of the $L^1(P)$-valued measure I_V. Since μ_V is σ-additive, I_V is a σ-additive $L^1(P)$-valued measure of bounded variation on Σ.

Every process $V \in SV$ is the difference of two increasing processes V^+ and V^- in SV, consequently all of the previous argument extends to an arbitrary process $V = (V_t) \in SV$, the variation of the $L^1(P)$-valued measure I_V, generated by the process $V = (V_t)$, being in the general case $|\mu_V|$, the variation of μ_V.

A stochastic process $Z = (Z_t)$ is integrable with respect to the measure I_V if and only if Z is integrable with respect to the stochastic process $V = (V_t)$ in the sense of 2.3. We denote the integral of $Z = (Z_t)$ with respect to the $L^1(P)$-valued measure I_V by $I_V(Z)$. In 2.3 the random variable $V(Z)_\infty$ was defined. The relation between $V(Z)_\infty$ and $I_V(Z)$ is that $V(Z)_\infty$ is a representative of the element
$$I_V(Z) \in L^1(P).$$

In particular the stochastic process $(V(Z)_t)$ defined in 2.3 is the unique right continuous modification of the $L^1(P)$-process defined by $t \to I_V(\chi_{[0,t] \times \Omega} Z)$, $(t \in R_+)$.

We shall meet this situation again, when we define the stochastic integral with respect to a larger class of stochastic processes. The main difference however will be that in the more general situation the stochastic integral only yields the $L^1(P)$-process, and not the stochastic process as in this section.

However we do not return to this more general situation until §11, because we first need to establish the prerequisite parts of the theory of stochastic processes.

§3 ADAPTED PROCESSES AND STOPPING TIMES

In this section we shall deal with a more special class of stochastic processes. The idea is, that at a time t the random variable X_t doesn't contain information about all of the σ-algebra F but only about a certain sub σ-algebra F_t of F. A random variable S with values in the parameter set T which is determined by the process up to $t = S$ will be called a stopping time. For a rigorous definition we need more notation.

Let (Ω, F, P) be a probability space and $(F_t)_{t \in T}$ $(T \subseteq \overline{R}_+)$ an increasing family of sub σ-algebras of F. A stochastic process $X = (X_t)_{t \in T}$ will be called F_t *adapted (or simply adapted)*, if for every $t \in T$ the random variable X_t is F_t-measurable.

If $T \to X_t$ is an $L^p(P)$-process $(1 \le p \le \infty)$, (X_t) is called adapted if X_t is an element of $L^p(F_t, P)$ for every $t \in T$, i.e. if X_t is invariant under the conditional expectation operator with respect to F_t for every $t \in T$.

We define the sub σ-algebras F_{t+} and F_{t-} of F_t by $F_{t+} = \wedge_{s>t} F_s$ and $F_{t-} = \vee_{s<t} F_s$. The family $(F_t)_{t \in T}$ is called *right continuous* if $F_{t+} = F_t$ holds for every $t \in T$ satisfying $\inf\{s : s > t, s \in T\} = t$. *Left continuity of* $(F_t)_{t \in T}$ is defined similarly.

If $X = (X_t)$ is a right continuous F_t-adapted process, then X is F_{t+}-adapted and $(F_{t+})_{t \in T}$ is a right continuous family of sub σ-algebras of F. Since we shall deal later on only with right continuous processes, we shall assume in future that the family $(F_t)_{t \in T}$ is right continuous. (Note that for example a Markov process X has a modification which is right continuous, provided very mild regularity conditions hold; (see for example [1] Chapter I). We shall furthermore assume that F is a complete σ-algebra and that F_o contains all P-null sets of F. Let us summarize in the following definition:

3.1 Definition: A quintuple $(\Omega, F, P, (F_t), T)$ will be called a *stochastic base*, if the following assertions are satisfied:

    (a) (Ω, F, P) is a complete probability space.

    (b) (F_t) $(t \in T)$ is a right continuous increasing family of sub σ-algebras of F, such that $\vee_{s \in T} F_s$ is equal to F.

    (c) F_o contains all null sets of F.

Note that we do not always need all the assumptions made about the stochastic base. The definition is however general enough for our purposes.

In 1.2 measurable stochastic processes were defined in terms of measurability with respect to the product σ-algebra $B \otimes F$, where B is the Borel σ-algebra on T. The following stronger form of measurability relative to a stochastic base implies that the stochastic process is adapted.

3.2 Definition: A stochastic process $X = (X_t)$ is called *progressively measurable* (with respect to the stochastic base $(\Omega, F, P, (F_t), T)$, if for every $t \in T$ the mapping $(s, \omega) \to X(s, \omega)$ on $(T \cap [o, t]) \times \Omega$ is measurable with respect to the product σ-algebra $B_t \otimes F_t$, where B_t is the Borel σ-algebra on $T \cap [o, t]$.

A subset $A \subseteq T \times \Omega$ is called *progressive* if the characteristic function χ_A of A is a progressively measurable process. The family of progressive sets form a σ-algebra Σ_o, and a process X is progressively measurable if and only if the function $(t, \omega) \to X(t, \omega)$ is measurable with respect to Σ_o.

It is clear that every progressively measurable process is adapted and measurable. The following theorem provides us with two classes of processes for which the converse is true.

3.3 Theorem: Let $X = (X_t)_{t \in R_+}$ be an F_t-adapted right continuous (resp. left continuous) stochastic process, then X is progressively measurable.

Proof: Suppose that X is right continuous and let an element t of R_+ be given. For every finite increasing sequence $\tau = (t_0 = 0, t_1, \ldots, t_{n-1}, t_n = t)$ of elements of the interval $[0,t]$, define the process X^τ as $X^\tau(s,\omega) = X(t_k, \omega)$ for $(s,\omega) \in \,]t_{k-1}, t_k] \times \Omega$ $(k = 1, \ldots, n)$. The function X^τ is $B_t \times F_t$-measurable for every τ, because X is adapted, and as $\sup\{t_k - t_{k-1} : t_k \in \tau, k = 1, \ldots, n\}$ tends to zero, X^τ tends to the restriction of X to $[0,t] \times \Omega$, hence X is progressively measurable.

The proof for a left continuous process X is similar.

3.4 Definition: Let (F_t) $(t \in T)$ be an increasing sequence of sub σ-algebras of a σ-algebra F. A function S on Ω with values in T is called a *stopping time*, if for every $t \in T$ the set $\{\omega \in \Omega : S(\omega) \leq t\}$ is an element of F_t. We shall write $\{S \leq t\}$ for the set $\{\omega \in \Omega : S(\omega) \leq t\}$ in future for the sake of brevity.

An immediate consequence of the definition is, that the supremum and infimum of a finite number of stopping times is again a stopping time. Moreover, if the supremum (resp. infimum) of a countable family of stopping times exists as a function from Ω to T, then this supremum (resp. infimum) is a stopping time.

For a stopping time S the *σ-algebra of events known at time S* is defined as
$$F_S = \{A \in F : A \cap \{S \leq t\} \in F_t \text{ for all } t \in T\}.$$

A simple example of stopping times are the functions S on Ω, which are constant equal to an element $t \in T$. For these stopping times we have $F_S = F_t$.

3.5 Proposition: Let S and S' be stopping times. The following assertions hold:

    (a) S is F_S-measurable.

    (b) If $S \leq S'$ then $F_S \subseteq F_{S'}$.

    (c) If A is an element of F_S then $A \cap \{S \leq S'\}$ is an element of $F_{S'}$.

Proof: The assertion (a) follows immediately from the definitions.
(b) Let B be an element of F_S and t an element of T. The set $B \cap \{S' \leq t\}$ is equal to the set $B \cap \{S \leq t\} \cap \{S' \leq t\}$, and since $\{S' \leq t\}$ and $B \cap \{S \leq t\}$ are elements of F_t, we conclude that $B \cap \{S' \leq t\}$ is an element of F_t. Assertion (c) finally follows from the equality

$$A \cap \{S \leq S'\} \cap \{S' \leq t\} = (A \cap \{S \leq t\}) \cap \{S' \leq t\} \cap \{S \wedge t \leq S' \wedge t\}.$$

Each of the three sets on the left hand side is F_t-measurable, the first one because A is an element of F_S, the second one because S' is a stopping time, and the third because $S \wedge t$ and $S' \wedge t$ are F_t-measurable random variables.

3.6 Definition: Let $X = (X_t)$ $(t \in T)$ be a stochastic process on a stochastic base $(\Omega, F, P, (F_t), T)$, and let S be an F_t-stopping time. We write X_S for the random variable $\omega \to X(S(\omega), \omega)$ and we define the process X, stopped at S, as $X^S = (X_{t \wedge S})$.

3.7 Theorem: If S is a stopping time and if $X = (X_t)$ is a progressively measurable process, then X_S is a F_S-measurable random variable and the stopped process X^S is progressively measurable.

Proof: In order to prove the first assertion, we have to show that for every Borel subset B of R (the state space of X) and for every $t \in T$ the set $\{X_S \in B\} \cap \{S \leq t\}$ is an element of F_t. However we have the equality $\{X_S \in B\} \cap \{S \leq t\} = \{X_{t \wedge S} \in B\} \cap \{S \leq t\}$, consequently it suffices to prove the second assertion of the theorem. Being a stopping time smaller than the constant stopping time t, $t \wedge S$ is a F_t-measurable random variable according to Proposition 3.5, hence the mapping $(s,\omega) \to (t \wedge S(\omega), \omega)$ is $B_t \otimes F_t$-measurable as a mapping on $(T \cap [o,t] \times \Omega)$, and consequently the mapping $(s,\omega) \to X(t \wedge S(\omega), \omega)$ is measurable as the composition of two measurable maps. This completes the proof of the second assertion of the theorem.

2 Generalized martingales and general theory of processes

§4   DISCRETE MARTINGALES

In this section we study a class of stochastic processes which play an important role in the theory of stochastic integration.

The first part of this section will mainly be concerned with convergence theorems and inequalities, in the second part vectorspaces of (discrete) martingales are studied. Here a crucial result will be proved, which will yield a sufficient condition for summability of a quasimartingale (see §9 and §11 for the definitions).

<u>4.1 Definition</u>:   Let $(\Omega, F, P, (F_t), T)$ be a stochastic base.   An $L^1(P)$-process $t \to X_t$ is called a *martingale*, if for every pair s,t of elements of T, $s \le t$ implies $E(X_t | F_s) = X_s$.

If in the last equality '=' is replaced by '≤' (resp. '≥'), the process $t \to X_t$ will be called a *supermartingale* (resp. *submartingale*).

If $t \to X_t$ is a norm bounded function from T to $L^p(P)$ ($1 \le p \le \infty$) and a (super, sub) martingale, the process will be called a *p-integrable* (super, sub) martingale. If the range of the function $t \to X_t$ is a uniformly integrable subset of $L^1(P)$, the process will be called *uniformly integrable*.

The following 'mean convergence theorem' applies to a situation much more general than the situation considered here. It will not be used in this generality in the sequel, it is stated here however, because it is a genuine convergence theorem for $L^p(P)$-processes, i.e. it is not concerned with the convergence of the paths of the process.

4.2 Theorem: Let (Ω, F, P) be a probability space, T an upwards filtering index set, and $(F_t)_{t \in T}$ an increasing family of sub σ-algebras of F.

If $(X_t)_{t \in T}$ is a family of elements X_t of $L^p(P)$ $(1 \leq p < \infty)$ which satisfies

(i) $s \leq t$ for two elements s and t of T implies $E(X_t | F_s) = X_s$

(ii) $\{X_t : t \in T\}$ is uniformly integrable in the case $p = 1$, $\{X_t : t \in T\}$ is bounded in $L^p(P)$ in case $p > 1$; then $(X_t)_{t \in T}$ converges to an element $X_\infty \in L^p(P)$ and $X_t = E(X_\infty | F_t)$ holds for all $t \in T$.

Proof: We prove the case $p = 1$, the proof for $p > 1$ is similar.

Being a uniformly integrable and hence relatively weakly compact set in $L^1(P)$, the net $(X_t)_{t \in T}$ has a weak cluster point $X_\infty \in L^1(P)$. Let us first show that $E(X_\infty | F_t) = X_t$ holds for $t \in T$. Suppose an $\varepsilon > 0$, $g \in L^\infty(P)$ and a $t \in T$ is given. There is a $s \geq t$ such that $|E[(X_\infty - X_s)E(g|F_t)]| \leq \varepsilon$ and hence

$$|E[(E(X_\infty | F_t) - X_t)g]| = |E[E(X_\infty - X_s | F_t)g]|$$
$$= |E[(X_\infty - X_s)E(g|F_t)]|$$
$$\leq \varepsilon$$

holds, i.e. the equality $E(X_\infty | F_t) = X_t$ is satisfied for every $t \in T$.

Furthermore, by virtue of the Hahn Banach theorem, X_∞ is an element of the (norm) closed convex hull of $\{X_t : t \in T\}$, hence for a given $\varepsilon > 0$ there are elements of T, $t_1, \ldots, t_n$, and real numbers $\beta_1, \ldots, \beta_n$ $(0 \leq \beta_i \leq 1, \Sigma_i \beta_i = 1)$, such that $\|X_\infty - \Sigma_i \beta_i X_{t_i}\|_1 \leq \varepsilon$ holds. Consequently for every $s \in T$ satisfying $s \geq t$, where t is an element of T majorizing $\{t_1, \ldots, t_n\}$, we have:

$$\|X_\infty - X_s\|_1 \leq \|X_\infty - \Sigma_i \beta_i X_{t_i}\|_1 + \|\Sigma_i \beta_i X_{t_i} - E(X_\infty | F_s)\|_1$$

$$\leq 2\|X_\infty - \Sigma_i \beta_i X_{t_i}\|_1$$
$$\leq 2\varepsilon .$$

This completes the proof of the theorem.

For the rest of the section a stochastic base $(\Omega, F, P, (F_n), N)$ will be fixed if nothing else is said. (N denotes the nonnegative integers).

It is clear that every $L^p(P)$-process $n \to X_n$ has a (up to indistinguishability) unique modification $X = (X_n)$. In the sequel we shall be concerned with properties of the paths of these modifications $X = (X_n)$ of the (super, sub) martingale $n \to X_n$. The processes $X = (X_n)$ are again called (super, sub) martingales, however the notation $n \to X_n$ will be exclusively used if we want to stress that the process is considered as an $L^p(P)$-process.

4.3 Definition: A stochastic process $Z = (Z_n)$ will be called *predictable* if Z_n is F_{n-1}-measurable for every $n \in N$. Let now $X = (X_n)$ be an adapted stochastic process and write ΔX_n for the difference
$$X_n = X_n - X_{n-1}, (X_{-1} = 0).$$
The *transform* $Y = X(Z)$ of the process X by $Z = (Z_n)$ is defined as $Y_n(\omega) = X(Z)_n(\omega) = \Sigma_{k=0}^{n} \Delta X_n(\omega) Z_n(\omega)$.

If Z is adapted, then $X(Z) = (X(Z)_n)$ is an adapted stochastic process, and if Z is predictable, we have

4.4 Theorem: Let $X = (X_n)$ be a supermartingale, $Z = (Z_n)$ a nonnegative predictable process.

If all the random variables $X(Z)_n$ are integrable, then $X(Z) = (X(Z)_n)$ is a supermartingale.

Proof: We have $E[X(Z)_{n+1} - X(Z)_n | F_n] = E[\Delta X_{n+1} Z_{n+1} | F_n] = Z_{n+1} E(\Delta X_{n+1} | F_n)$

for every $n \in \mathbb{N}$, the assertion thus follows from the definition of a supermartingale.

Corollary 1: Let $X = (X_n)$ be a martingale, $Z = (Z_n)$ a predictable process. If the random variables $X(Z)_n$ $(n \in \mathbb{N})$ are integrable, the transform $X(Z) = (X(Z)_n)$ is a martingale.

Corollary 2: Let $X = (X_n)$ be a supermartingale, S a stopping time. The stopped process X^S is a supermartingale.

Proof: Define the process $Z = (Z_n)$ as $Z_n = X_{\{n \leq S\}}$. Z is predictable and we have $X_n^S = X(Z)_n$ $(n \in \mathbb{N})$. The integrability condition follows from the inequality $|X_n^S| \leq \sum_{k=1}^{n} |X_k|$, this completes the proof of the Corollary in view of the previous theorem.

Corollary 3: Let $X = (X_n)$ be a suprermartingale and S, T a pair of bounded stopping times, satisfying $S \leq T$. X_S and X_T are integrable and the inequality $E(X_T | F_S) \leq X_S$ holds.

Proof Define the predictable process $Z = (Z_n)$ as $Z_n = X_{\{S < n \leq T\}}$. Z satisfied the hypothesis of Theorem 4.4, consequently $Y = X(Z)$ is a supermartingale, satisfying $Y_0 = 0$. If k is an integer, greater than $S \vee T$, then $Y_k = X_T - X_S$ holds, and we have $0 = E(Y_0) \geq E(Y_k) = E(X_T - X_S)$ We get the conditional form of this inequality, using a technique, which will be used frequently in the sequel: Let A be an element of F_S. Define the stopping times S' and T' as $S' = X_A S + X_{A^c} k$ and $T' = X_A T + X_{A^c} k$. S' is smaller than T', consequently $E[X_A(X_T - X_S)] = E(X_{T'} - X_{S'}) \leq 0$ holds for all $A \in F_S$, which proves the inequality

$$E(X_T - X_S | F_S) \leq 0.$$

The main tool for studying the behavior of the paths of a supermartingale is the 'Upcrossing Lemma', which is due to Doob.

Let $X = (X_n)$ be an adapted stochastic process, and a,b a pair of real numbers, satisfying $a < b$. We define a sequence (T_k) of stopping times by induction:

$$T_0(\omega) = \inf\{n : X_n(\omega) < a\}$$
$$T_1(\omega) = \inf\{n : n > T_0(\omega) \text{ and } X_n(\omega) > b\}$$

$$T_{2k}(\omega) = \inf\{n : n > T_{2k-1}(\omega) \text{ and } X_n(\omega) < a\}$$
$$T_{2k+1}(\omega) = \inf\{n : n > T_{2k}(\omega) \text{ and } X_n(\omega) > b\}$$

for $k \geq 1$, the infimum of the empty set being defined as ∞. T_{2k-1} is the k-th time an upcrossing of $[a,b]$ is completed.

Similarly T_{2k} is the k-th time a downcrossing of $[a,b]$ is completed.

Now put $U_a^b(\omega) = \sup\{k : T_{2k-1}(\omega) < \infty\}$. U_a^b is a random variable and gives the total number of completed upcrossings of $[a,b]$ by the path $X(\omega)$ of the process X.

The following lemma estimates the probability that more than p upcrossings take place in the case that X is a supermartingale.

<u>4.5 Doob's Upcrossing Lemma</u>: Let $X = (X_n)$ be a supermartingale, stopped at a bounded stopping time S.

Then $P(\{U_a^b > p\}) \leq 1/(b-a) \int_{\{U_a^b = p\}} (X_S - a)^- dP$ holds for every $p \in \mathbb{N}$ and every pair of real numbers a,b such that $a < b$.

<u>Proof</u>: We may assume that a is equal to 0. Define the stopping times V and W by

$V = T_{2p} \wedge S$ and $W = T_{2p+1} \wedge S$.

We have the inequality (4.5.1): $bP(\{U_0^b > p\}) \leq \int_{\{U_0^b > p\}} X_W^- dP \leq \int_{\{T_{2p} < \infty\}} X_W^+ dP$.

We now apply 4.4 Corollary 3 to the bounded stopping times V and W. Since the process X is stopped at S, the set $\{T_{2p} < \infty\}$ belongs to $F_{T_{2p}}$ and to F_S and hence to F_V. The fact that $X_V \leq o$ holds on the set $\{T_{2p} < \infty\}$ and 4.4 Corollary 3 yield $\int_{\{T_{2p} < \infty\}} X_W dP \leq \int_{\{T_{2p} < \infty\}} X_V dP \leq o$, which gives the estimate (4.5.2):

$$\int_{\{T_{2p} < \infty\}} X_W^+ dP \leq \int_{\{T_{2p} < \infty\}} X_W^- dP = \int_{\{T_{2p} < \infty, X_W < o\}} X_W^- dP$$

$$\leq \int_{\{T_{2p} < \infty, X_W < o\}} X_S^- dP .$$

In view of 4.4 Corollary 3, the last inequality follows from the facts that the process $(-X_n^-)$ is a supermartingale and the set $\{T_{2p} < \infty, X_W < o\}$ is an element of F_W.

On the other hand, if the $(p + 1)$th upcrossing is completed for a path $X(\omega)$, we must have $X_W(\omega) \geq b$, therefore on the set $\{T_{2p} < \infty, X_W < o\}$ a $(p + 1)$th upcrossing will never take place, i.e. U_0^b is equal to p on the set $\{T_{2p} < \infty, X_W < o\}$. Consequently the estimate (4.5.3)

$$\int_{\{T_{2p} < \infty, X_W < o\}} X_S^- dP \leq \int_{\{U_a^b = p\}} X_S^- dP$$

holds, and putting (4.5.1), (4.5.2) and (4.5.3) together, we finally get the desired inequality.

<u>Corollary 1</u>: For a supermartingale $X = (X_n)$, stopped at a bounded stopping time S we have the estimate:

$$E(U_a^b) \leq 1/(b - a) E[(X_S - a)^-] .$$

Proof: We have

$$E(U_a^b) = \Sigma_{k=0}^{\infty} P(\{U_a^b > k\})$$

$$\leq 1/(b-a)\Sigma_{k=0}^{\infty} \int_{\{U_a^b=k\}} (K_S - a)^- dP \leq 1/(b-a)E[X_S - a)^-].$$

Doob's Upcrossing Lemma will be used in the proof of the following.

4.6 Theorem: Let $X = (X_n)$ be a supermartingale, satisfying $\sup_n E(X_n^-) < \infty$. The sequence (X_n) converges P-almost surely to an integrable random variable X_∞.

Remark: Because of the inequality
$$\|X_n\|_1 = E(X_n^+ + X_n^-)$$
$$= E(X_n) + 2 E(X_n^-)$$
$$\leq E(X_0) + 2 E(X_n^-)$$

the condition $\sup_n E(X_n^-) < \infty$ is equivalent to $\sup_n \|X_n\|_1 < \infty$ for a supermartingale $X = (X_n)$.

Proof: Let a pair a,b of real numbers $a < b$ be given. Define an increasing sequence $(U_a^b(k))$ of positive random variables, where $U_a^b(k)$ ($k \in \mathbb{N}$) is the random variable of completed upcrossings of $[a,b]$ by the supermartingale X^k, i.e. the supermartingale X, stopped at the constant stopping time k. The sequence $(U_a^b(k))$ converges to U_a^b, the random variable of completed upcrossings of $[a,b]$ by X, hence we have

$$E(U_a^b) = \lim_k E(U_a^b(k))$$

and, since according to 4.5 Corollary 1 for every k

$$E(U_a^b(k)) \leq 1/(b-a)E[(X_k - a)^-] \leq 1/(b-a)[|a| + E(X_k^-)]$$

holds, we get

$$E(U_a^b) \leq 1/(b-a)[|a| + \sup_k E(X_k^-)] < \infty .$$

This shows that U_a^b is finite, P almost surely. On the other hand the set $\{U_a^b < \infty\}$ contains the set $H_{a,b}$ of all $\omega \in \Omega$ such that $\limsup_n X_n(\omega)$ is smaller than a. Hence $H_{a,b}$ is a P-null set and so is the union of all sets $H_{a,b}$, where a,b runs through all pairs of rationals with $a < b$.

This union however is just the set $\{\limsup_n X_n \neq \liminf_n X_n\}$, i.e. the set where (X_n) doesn't converge.

This completes the proof of the first part of the theorem, the integrability of the limit follows from Fatou's lemma and the inequality mentioned in the remark preceding the proof.

<u>Corollary 1</u>: Let $X = (X_n)$ be a uniformly integrable martingale (resp. supermartingale). The sequence (X_n) converges to a random variable X_∞ P-almost surely and in the $L^1(P)$-norm, and $X_n = E(X_\infty|F_n)$ (resp. $X_n \geq E(X_\infty|F_n)$) holds for all $n \in \mathbb{N}$.

As another application of the Upcrossing Lemma, Doob's inequalities will be proved. The 'Maximal inequality' will be divided into two parts.

<u>4.7 Lemma</u>: Let $X = (X_n)$ be a supermartingale. For every $\lambda \geq 0$ $\lambda P(\sup_n X_n \geq \lambda) \leq c \sup_n \|X_n\|_1$ holds, where the constant c can be taken equal to 1 if X is positive and equal to 2 otherwise.

<u>Proof</u>: Define T as $T(\omega) = \inf\{n : X_n(\omega) \geq \lambda\}$ and define the sequence (T_k) of bounded stopping times by $T_k = T \wedge k$. By 4.4 Corollary 3 $E(X_{T_k}) \leq E(X_0)$ holds for every k. If $X_n(\omega) \geq \lambda$ holds for some $n \leq k$, we have $X_{T_k}(\omega) \geq \lambda$, otherwise $X_{T_k}(\omega) = X_k(\omega)$. Accordingly we have

$$E(X_o) \geq \lambda P(\{\sup_{n \leq k} X_n \geq \lambda\}) + \int_{\{\sup_{n \leq k} X_n < \lambda\}} X_k dP$$

and for k tending to infinity

$$\lambda P(\{\sup_n X_n \geq \lambda\}) \leq E(X_o) + \sup_n \int X_n^- dP \leq 2 \sup_n \|X_n\|_1 .$$

4.8 Lemma: Let $X = (X_n)$ be a supermartingale. For every $\lambda > 0$

$$\lambda P(\{\inf_n X_n \leq -\lambda\}) \leq \sup_n E(X_n^-)$$

holds.

Proof: Define $T(\omega) = \inf\{n : X_n(\omega) \leq -\lambda\}$ and define the sequence (T_k) of bounded stopping times by $T_k = T \wedge k$. By 4.5 Corollary 3 $E(X_{T_k}) \geq E(X_k)$ holds for every k and as in the proof of Lemma 4.7 we get

$$E(X_k) \leq -\lambda P(\{\inf_{n \leq k} X_n \leq -\lambda\}) + \int_{\{\inf_{n \leq k} X_n > -\lambda\}} X_k dP .$$

This yields the inequality (4.8.1):

$$\lambda P(\{\inf_{n \leq k} X_n \leq -\lambda\}) \leq E(-X_k) + \int_{\{\inf_{n \leq k} X_n > -\lambda\}} X_k dP$$

$$= \int_{\{\inf_{n \leq k} X_n \leq -\lambda\}} -X_k dP$$

$$\leq E(X_k^-) .$$

Thus the result follows as k tends to infinity.

From Lemma 4.8 we conclude

4.9 Doob's Maximal Theorem: Let $X = (X_n)$ be a martingale. The inequality

$$\lambda P(\{\sup_n |X_n| \geq \lambda\}) \leq \sup_n \|X_n\|_1$$

holds for every positive real number $\lambda > 0$.

Proof: The proof follows immediately from the fact that $n \to -|X_n|$ is a negative supermartingale.

For supermartingale lemmas 4.7 and 4.8 yield

4.10 Theorem: Let $X = (X_n)$ be a supermartingale. The inequality

$$P(\{\sup_n |X_n| \geq \lambda\}) \leq 3 \sup_n \|X_n\|_1$$

holds for every positive real number $\lambda > 0$.

The following maximal theorem applies to positive p-integrable ($1 < p \leq \infty$) submartingales and therefore with the obvious modifications to p-integrable martingales. The case $p = 1$, to which this theorem cannot be extended, will be treated in the last part of this section.

4.11 Theorem: Let $X = (X_n)$ be a positive submartingale and put $X^* = \sup_n X_n$. For $1 < p \leq \infty$ X^* is p-integrable if and only if $\sup_n \|X_n\|_p < \infty$ holds. Moreover the inequality $\|X^*\|_p \leq p/p-1 \sup_n \|X_n\|_p$ is satisfied for $1 < p < \infty$.

Proof: Put $X_k^* = \sup_{n \leq k} X_n$. The first assertion of the theorem is obvious for $p = \infty$, in view of the Monotone Convergence Theorem it therefore suffices to prove that the inequality $\|X_k^*\|_p \leq p/p-1 \sup_{n \leq k} \|X_n\|_p$ holds for every k ($1 < p < \infty$).

Taking into account Lemma 4.8, this inequality is an immediate consequence of the Marcinkiewicz Interpolation Theorem (see e.g. [30]). However, using the inequality (4.8.1), a more elementary proof can be given.

According to (4.8.1) we have $\lambda P(\{X_k^* \geq \lambda\}) \leq \int_{\{X_k^* \geq \lambda\}} X_k \, dP$, and the result follows from

4.12 Lemma: Let f and g be two positive random variables, satisfying $P(\{f \geq \lambda\}) \leq \int_{\{f \geq \lambda\}} g \, dP$ for every $\lambda > 0$. The inequality $\|f\|_p \leq p/p-1 \|g\|_p$ holds for $1 < p < \infty$.

Proof (of the lemma): Let F be an increasing continuous function on the positive real line R_+ satisfying $F(o) = o$, and let $D(\lambda) = P(\{f \geq \lambda\})$ the distribution function of f. Integration by parts yields

$$E(F \circ f) = -\int_0^\infty F(\lambda)dD(\lambda) \leq \int_0^\infty D(\lambda)dF(\lambda).$$

By hypothesis we have $D(\lambda) \leq 1/\lambda \int_{\{f \geq \lambda\}} g \, dP$ and we get by virtue of Fubini's theorem

$$E(F \circ f) \leq \int_0^\infty [1/\lambda \int_{\{f \geq \lambda\}} g \, dP] dF(\lambda)$$

$$= E[g \int 1/\lambda \, \chi_{\{f \geq \lambda\}} dF(\lambda)]$$

This yields for $F(\lambda) = \lambda^p$

$$E(f^p) \leq p/p-1 \, E(gf^{p-1}) \leq p/p-1 \|g\|_p \|f^{p-1}\|_{p/p-1}$$

$$= p/p-1 \|g\|_p (\|f\|_p)^{p-1}.$$

If $\|f\|_p$ is finite, the desired inequality follows immediately, and since for every n the random variable $f \wedge n$ satisfies the hypothesis of the lemma, we get the result by letting n tend to infinity.

In Theorem 4.6 regularity properties for integrable supermartingales were established. In order to prove a generalization of 4.4 Corollary 3, namely the optional sampling theorem, we are going to establish this regularity for reverse supermartingales.

<u>4.13 Definition</u>: Let (Ω, F, P) be a probability space, (F_n) $(n=0,1,2,\ldots)$ a *decreasing* sequence of sub σ-algebras of F. An $L^1(P)$-process $X = (X_n)$ is called a *reverse supermartingale* (with respect to (F_n)), if the inequality $E(X_m|F_n) \leq X_n$ holds for every pair of nonnegative integers satisfying $m \leq n$. Reverse martingales and submartingales are defined analogously.

4.14 Proposition: Let $X = (X_n)$ be a reverse supermartingale with respect to the decreasing sequence of σ-algebras (F_n). Then either $\lim_{n \to \infty} E(X_n)$ is equal to $+\infty$ or the range of the sequence (X_n) forms a uniformly integrable subset of $L^1(P)$.

Proof: $E(X_n)$ is an increasing function of n, hence the limit

$$\beta = \lim_{n \to \infty} E(X_n)$$

exists. Suppose β is finite. Since the set $\{E(X_0|F_n) : n = 0,1,2,\ldots\}$ is uniformly integrable, it suffices to prove that the reverse supermartingale $n \to X_n - E(X_0|F_n)$ is uniformly integrable. We therefore may assume without loss of generality that X_n is positive for every n. Let an $\varepsilon > 0$ be given and choose a k large enough, so that $\beta - E(X_k)$ is smaller than $\varepsilon/2$. For $n \geq k$ we deduce from the supermartingale inequality, λ being a positive real number

$$\int_{\{X_n > \lambda\}} X_n \, dP = E(X_n) - \int_{\{X_n \leq \lambda\}} X_n \, dP$$

$$\leq E(X_n) - \int_{\{X_n \leq \lambda\}} X_k \, dP$$

$$= E(X_n) - E(X_k) + \int_{\{X_n > \lambda\}} X_k \, dP$$

$$\leq \varepsilon/2 + \int_{\{X_n > \lambda\}} X_k \, dP .$$

Now $P(\{X_n > \lambda\}) \leq E(X_n)/\lambda \leq \beta/\lambda$ holds, hence there is a λ_0 such that $\int_{\{X_n > \lambda_0\}} X_k \, dP \leq \varepsilon/2$ holds for all n. This completes the proof of the proposition.

From Proposition 4.14 we can deduce the following convergence theorem, for which the proof is similar to the proof of Theorem 4.6 and will be left to the reader.

4.15 Theorem: Let $X = (X_n)$ be a reverse supermartingale with respect to the decreasing sequence of σ-algebras (F_n), such that $\lim_{n \to \infty} E(X_n) < \infty$ holds. The sequence (X_n) of random variables converges almost surely and in the $L^1(P)$ - norm to an integrable random variable X_∞.

If $X = (X_n)$ is a reverse martingale, $E(X_n) = E(X_0)$ is satisfied for all n and we have $X_\infty = E(X_0 | \cap_n F_n)$.

We are now able to prove the following theorem:

4.16 Doob's Optional Sampling Theorem: Let $X = (X_n)$ be a supermartingale with respect to the stochastic base $(\Omega, F, P, (F_n), \bar{N})$, where $\bar{N}$ is equal to $N \cup \{\infty\}$ and F_∞ is equal to the σ-algebra $\vee_{n \in N} F_n$.

If S and T are two stopping times with $S \leq T$, then X_S and X_T are integrable random variables, and the inequality $E(X_T | F_S) \leq X_S$ holds. If $X = (X_n)$ is a martingale we have $X_S = E(X_T | F_S) = E(X_\infty | F_S)$

Proof: Put $Y_n = E(X_\infty | F_n)$ and $Z_n = X_n - Y_n$. $Y = (Y_n)$ is a uniformly integrable martingale and $Z = (Z_n)$ is a positive supermartingale.

Because X is the sum of Y and Z, it will be sufficient to prove the theorem for Y and Z separately.

Let us first consider Y. Define the stopping times S_n by $S_n(\omega) = S(\omega)$ on $\{S \leq n\}$ and $S_n(\omega) = \infty$ on $\{S > n\}$. The finite sequence $\{0, 1, \ldots, n, \infty\}$ is order isomorphic to $\{0, 1, \ldots, n, n+1\}$ and by 4.4 Corollary 3 we have (4.16.1): $Y_{S_n} = E(Y_\infty | F_{S_n})$. The random variables Y_S and Y_{S_n} coincide on the set $\{S \leq n\} \cup \{S = \infty\}$, hence the left hand side of (4.16.1) converges to Y_S as n tends to ∞. Because of $F_S = \cap_{n \leq 1} F_{S_n}$, Theorem 4.15 yields

$\lim_{n\to\infty} E(Y_\infty | F_{S_n}) = E(Y_\infty | F_S)$, i.e. we have $Y_S = E(Y_\infty | F_S)$. Finally the equality $Y_S = E[E(Y_\infty | F_T) | F_S] = E(Y_T | F_S)$ proves the second assertion of the theorem.

Let us now prove the first assertion of the theorem for the positive supermartingale $Z = (Z_n)$. Define S_n and T_n as in the first part of the proof. (Z_{S_n}) and (Z_{T_n}) are increasing sequences of positive random variables because of $Z_\infty = 0$, which converge almost surely to Z_S resp. to Z_T. From the inequalities $E(Z_0) \geq E(Z_{S_n})$ resp. $E(Z_0) \geq E(Z_{T_n})$ for all n and the Monotone Convergence Theorem we conclude that Z_S and Z_T are integrable and that the convergence takes place in the $L^1(P)$-norm topology.

Moreover we have according to 4.4 Corollary 3 $Z_{S_n} \geq E(Z_{T_n} | F_{S_n})$ for every n, hence Theorem 4.15 yields as in the first part of the proof

$$Z_S = \lim_{n\to\infty} Z_{S_n} \geq \lim_{n\to\infty} E(Z_{T_n} | F_{S_n}) = \lim_{n\to\infty} E(Z_{T_n} | F_S)$$

$$= E(Z_T | F_S).$$

<u>4.17</u>: The family M of martingales forms a vector space under the operations $\alpha F + \beta G = (\alpha F_n + \beta G_n)$ ($\alpha, \beta \in R$; $F = (F_n)$ and $G = (G_n)$ martingales). Given an element f of $L^p(P)$ ($1 \leq p \leq \infty$), the range of the sequence (F_n), defined by $F_n = E(f | F_n)$, forms a martingale and according to Theorem 4.2 every uniformly integrable martingale is of this form for some $f \in L^1(P)$. The mapping $f \to F = (F_n)$ defined above is an injective linear map from $L^p(P)$ onto the subspace M^p of M of p-integrable martingales in the case $1 < p \leq \infty$ or. of uniformly integrable martingales in the case $p=1$. This induces a natural Banach space structure on M^p by means of the norm $\|F\|_p = \|\lim_n F_n\|_p = \|F_\infty\|_p$. According to Theorem 4.11 the norm $\|F\| = \|\sup_n |F_n|\|_p$ is equivalent to the norm $\|F\|_p$ defined above for $p > 1$. This result doesn't

hold for $p = 1$, as the following example shows:

Let Ω be the set of natural numbers $\{1,2,\ldots\}$, F the σ-algebra of all subsets of Ω and P a probability measure on F such that the measure $P(A_k) = a_k$ of the set $A_k = \{k+1, k+2, \ldots\}$ is strictly greater than zero for all $k = 0, 1, \ldots$ Let F_n $(n=0,1,\ldots)$ be the σ-algebra generated by the sets $A_0, A_1, \ldots, A_n$. Put $b_k = 1/k^2$, $s_k = \Sigma_{m=k}^{\infty} b_m$ and $f = \Sigma_{k=1}^{\infty} a_k^{-1} b_k X_{A_k}$. f is integrable and consequently and corresponding martingale $F = (F_n)$ $(F_n = E(f|F_n))$ is uniformly integrable. On the other hand we have

$$\sup_{n \leq m} |F_n| \geq \Sigma_{k=1}^{m} a_k^{-1} s_k X_{A_k},$$ i.e. $\sup_n |F_n|$ is not integrable.

The remainder of this section is devoted to the study of the subspace $H^1 = \{F \in M^1 : E(\sup_n |F_n|) < \infty\}$ of M^1.

We start by comparing equivalent norms on $L^1(P)$. In order to make the notation less cumbersome, the following abbreviations are introduced: We write L^1 for $L^1(P)$ and L^∞ for $L^\infty(P)$. Furthermore if $F_1 \subseteq F_2 \subseteq \ldots \subseteq F_n = F$ is a finite increasing sequence of sub σ-algebras of F, we write E^k for the conditional expectation operator $f \to E(f|F_k)$ ($1 \leq k \leq n$), $E^0 = 0$ and $E_j^k = E^k - E^j$.

On the product space $(L^1)^n$ we define two equivalent norms $\bar{p}_1$ and $\bar{p}_2$ by

$$\bar{p}_1(f_k) = \|\sup_k |f_k|\|_1 \text{ and}$$

$$\bar{p}_2(f_k) = \|(\Sigma_k f_k^2)^{1/2}\|_1, \text{ where } (f_k) \text{ is an element of } (L^1)^n.$$

The dual of $(L^1)^n$ can be identified as $(L^\infty)^n$ under the dual bilinear form $\langle (f_k), (g_k) \rangle = E(\Sigma_k f_k g_k)$ $((f_k) \in (L^1)^n, (g_k) \in (L^\infty)^n)$. The respective dual norms are

$$\bar{q}_1(g_k) = \|\Sigma_k |g_k|\|_\infty \quad \text{and}$$

$$\bar{q}_2(g_k) = \|(\Sigma_k g_k^2)^{1/2}\|_\infty \,.$$

L^1 can be imbedded into $(L^1)^n$ by means of the linear maps

$$\rho_1 : f \to (E^k f) \quad \text{and}$$

$$\rho_2 : f \to (E^k_{k-1} f) \,.$$

If L^1 is considered as a subspace of $((L^1)^n, \bar{p}_i)$ by means of ρ_i ($i = 1,2$), we shall write p_i for the corresponding norms on L^1, i.e. we have for an element $f \in L^1$

$$p_1(f) = \bar{p}_1(E^k f) = \|\sup_k |E^k f|\|_1 \quad \text{and}$$

$$p_2(f) = \bar{p}_2(E^k_{k-1} f) = \|(\Sigma_k (E^k_{k-1} f)^2)^{1/2}\|_1$$

and the norms p_1 and p_2 are equivalent to the L^1-norm $\|f\|_1$.

This yields for the corresponding dual norms q_i ($i=1,2$) on L^∞

$$q_i(g) = \inf \{\bar{q}_i(g_k) : (g_k) \in (L^\infty)^n, \ \rho_i'(g_k) = g\},$$

where the adjoints ρ_i' of ρ_i are given by

$$\rho_1'(g_k) = \Sigma_k E^k g_k \quad \text{and} \quad \rho_2'(g_k) = \Sigma_k E^k_{k-1} g_k.$$

The norms q_1 and q_2 are equivalent to the L^∞-norm $\|g\|_\infty$ and we define a third equivalent norm q_3 on L^∞ by

$$q_3(g) = \sup_k \|(E^k[(E^n_{k-1} g)^2])^{1/2}\|_\infty$$

$$= \sup_k \|(E^k[(g - E^{k-1} g)^2])^{1/2}\|_\infty \,.$$

The following proposition shows, that the norms q_1, q_2 and q_3 are not only equivalent, but comparable using constants independant of the sequence of σ-algebras $F_1, \ldots, F_n$.

4.18 Proposition: The norms q_i ($i = 1, 2, 3$) on L^∞ satisfy the following inequalities:

a) $q_3 \leq \sqrt{5}\, q_1$

b) $q_1 \leq 5\, q_2$

c) $q_2 \leq \sqrt{2}\, q_3$

Proof: a) It suffices to show that for every element (g_k) of $(L^\infty)^n$ with $\bar{q}_1(g_k) \leq 1$ the inequality $E^j[(E^n_{j-1}\rho'_1(g_k))^2] \leq 5$ holds for $j = 1, 2, \ldots n$. Let (g_k) be given as above. We have

$$E^n_{j-1}\rho'_1(g_k) = E^n_{j-1}(\Sigma_k E^k g_k) = E^n_{j-1}(\Sigma^n_{k=j} E^k g_k).$$

Putting $\Sigma^n_{k=j} E^k g_k = h_j$, we get

$$E^j[(E^n_{j-1}h_j)^2] = E^j(h_j^2) - 2\, E^j h_j E^{j-1} h_j + (E^{j-1}h_j)^2$$

$$\leq E^j(h_j^2) + 3$$

because for $i \leq j$ $|E^i(h_j)| \leq E^i(\Sigma^n_{k=j}|g_k|) \leq 1$ holds. For the same reason we get

$$E^j(h_j^2) \leq 2\, \Sigma^n_{k=j} \Sigma^n_{m=k} E^j[(E^k|g_k|)(E^m|g_m|)]$$

$$= 2\, \Sigma^n_{k=j} \Sigma^n_{m=k} E^j[(E^k|g_k|)(E^k|g_m|)]$$

$$= 2\, \Sigma^n_{k=j} E^j[(E^k|g_k|)(\Sigma^n_{m=k}|g_m|)]$$

$$\leq 2\, \Sigma^n_{k=j} E^j(E^k|g_k|)$$

$$\leq 2$$

Together with the inequality above this yields $E^j[(E^n_{k-1}h_j)^2] \leq 5$, i.e. $[\bar{q}_3(g_k)]^2 \leq 5$, the desired result. Next we prove c). We shall show that for every $f \in L^1$ and $g \in L^\infty$ the inequality $E(fg) \leq \sqrt{2}\, p_2(f)\, q_3(g)$

holds. (This inequality is called **Fefferman's inequality**). Define $S_j = S_j(f) = (\sum_{k=1}^{j}(E_{k-1}^{k}f)^2)^{1/2}$ $(j = 1,2,\ldots,n)$, so that $p_2(f)$ becomes $p_2(f) = \|S_n\|_1$.

By Schwartz's inequality we have

$$E(fg) = E[\sum_{j=1}^{n}(E_{j-1}^{j}f)(E_{j-1}^{j}g)]$$

$$= E[\sum_{j=1}^{n}(E_{j-1}^{j}f)S_j^{-1/2}(E_{j-1}^{j}g)S_j^{1/2}]$$

$$\leq E[(\sum_{j=1}^{n}(E_{j-1}^{j}f)^2 S_j^{-1})^{1/2} (\sum_{j=1}^{n}(E_{j-1}^{j}g)^2 S_j)^{1/2}].$$

Putting $S_0 = 0$, this yields according to Holder's inequality $E(fg) \leq A^{1/2}B^{1/2}$ where $A = \sum_{j=1}^{n} E[(S_j^2 - S_{j-1}^2)S_j^{-1}]$ and $B = \sum_{j=1}^{n} E[(E_{j-1}^{j}g)^2 S_j]$. Since $S_j + S_{j-1} \leq 2 S_j$, we get

$$A \leq 2 \sum_{j=1}^{n} E(S_j - S_{j-1}) = 2 E(S_n) = 2 p_2(f)$$

and for B the following estimate holds:

$$B = \sum_{j=1}^{n} E[(E_{j-1}^{j}g)^2 (\sum_{l=1}^{j} S_l - S_{l-1})]$$

$$= \sum_{l=1}^{n} E[(S_l - S_{l-1}) \sum_{j=l}^{n}(E_{j-1}^{j}g)^2]$$

$$= \sum_{l=1}^{n} E[(S_l - S_{l-1}) E^l[(E_{l-1}^{n}g)^2]]$$

$$\leq [q_3(g)]^2 p_2(f).$$

Thus we get finally $E(fg) \leq \sqrt{A}\sqrt{B} \leq \sqrt{2}\, p_2(f)\, q_3(g)$.

In order to prove b) we shall prove the corresponding inequality in the predual L^1 of L^∞, i.e. the inequality $p_2 \leq 5 p_1$.

For an element $f \in L^1$ define $f_* = \sup_k |E^k f|$, and $S = S(f)$ by $S = (\sum_{j=1}^{n}(E_{j-1}^{j}f)^2)^{1/2}$, so that $E(f_*) = p_1(f)$ and $E(S) = p_2(f)$ holds.

The random variable $g = \sum_{j=1}^{n}(E^{j-1}f)(E_{j-1}^{j}f_*^{-1})$ satisfies

$$E^j[(E^n_{j-1}g)^2] = E^j[\Sigma^n_{k=j}(E^k_{k-1}g)^2]$$

$$= E^j[\Sigma^n_{k=j}(E^{k-1}f)^2(E^k_{k-1}f_*^{-1})^2]$$

$$= E^j[\Sigma^n_{k=j+1}(E^{k-1}f)^2((E^k f_*^{-1})^2 - (E^{k-1}f_*^{-1})^2)]$$

$$+ (E^{j-1}f)^2 (E^j_{j-1}f_*^{-1})^2$$

$$\leq E^j[f_*^2(E^n f_*^{-1})^2] + [(E^{j-1}f)(E^j f_*^{-1} \vee E^{j-1}f_*^{-1})]^2$$

$$\leq 2$$

for every $j = 1,2,\ldots,n$, i.e. g is an element of L^∞ and $q_3(g)$ is smaller than $\sqrt{2}$.

Using this result and the inequality c), which is already proved, we get

$$|E[\Sigma^n_{j=1} (E^j_{j-1}f)(E^{j-1}f) f_*^{-1}]| =$$

$$= |E[\Sigma^n_{j=1} f (E^{j-1}f)(E^j_{j-1}f_*^{-1})]| = |E(fg)| \leq \sqrt{2}\ p_2(f)\ q_3(g)$$

$$\leq 2\ p_2(f).$$

Taking into account that $S^2 = (E^n f)^2 - 2\Sigma^n_{j=1}(E^j_{j-1}f)(E^{j-1}f)$ holds, we conclude by Holder's inequality and by the inequality just proved

$$(p_2(f))^2 = (E(S))^2 \leq E(f_*)\ E(S^2 f_*^{-1})$$

$$\leq p_1(f)[E(f_*) - 2\ E(\Sigma^n_{j=1}(E^j_{j-1}f)(E^{j-1}f)\ f_*^{-1})]$$

$$\leq (p_1(f))^2 + 4\ p_2(f)\ p_1(f)$$

or equivalently

$$(p_2(f) - 2\ p_1(f))^2 \leq 5\ (p_1(f))^2.$$

This gives finally $p_2(f) \leq (2 + \sqrt{5})\ p_1(f) \leq 5\ p_1(f)$.

__Corollary 1:__ The norms p_1 and p_2 on $L^1(P)$ satisfy $p_1 \leq \sqrt{10}\, p_2$ and $p_2 \leq 5\, p_1$.

The following corollary will be used in §12 in connection with continuous parameter martingales. For the discrete case it implies that, given a bounded predictable process $Z = (Z_n)$, the transform $X \to X(Z)$ (see Definition 4.3) is a continuous linear operation on the space of martingales H^1.

__Corollary 2:__ Let $\beta_1, \ldots, \beta_n$ be a sequence of real numbers satisfying $|\beta_i| \leq 1$ for $i = 1, \ldots, n$. For an element f of $L^1(P)$ define $\sum_{i=1}^{n} \beta_i\, E_{i-1}^i f$. Then $p_1(h) \leq 16\, p_1(f)$ and consequently $\|h\|_1 \leq 16\, p_1(f)$ hold.

__Proof:__ We have $p_2(h) = E(\sum_{i=1}^{n} \beta_i^2 (E_{i-1}^i f)^2)^{1/2} \leq p_2(f)$, and hence by Corollary 1 $p_1(h) \leq \sqrt{10}\, p_2(h) \leq \sqrt{10}\, p_2(f) \leq \sqrt{250}\, p_1(f)$

$$\leq 16\, p_1(f)$$

Let us now return to the space $H^1 = \{F = (F_n) \in M_1 : E(\sup_n |F_n|) < \infty\}$ of martingales $F = (F_n)$ which are order bounded in $L^1(P)$. In view of 4.18 Corollary 1 we have the following.

__4.19 Theorem:__ The space H^1 is a Banach space under the equivalent norms

$$p_1(F) = E(\sup_n |F_n|) \quad \text{and}$$

$$p_2(F) = \sup_n E[(\sum_{j=1}^{n} (F_{j+1} - F_j)^2 + F_0^2)^{1/2}]$$

More precisely, p_1 and p_2 are related by $p_1 \leq \sqrt{10}\, p_2$ and $p_2 \leq 5\, p_1$.

It is now easy to determine the dual of H^1. As usual we identify martingales $F = (F_n) \in H^1$ with the elements $f \in L^1(P)$, defined by $f = \lim_n F_n$. First notice that according to Theorem 4.9 (Doob's Maximal

Theorem) and the Dominated Convergence Theorem for every $f \in H^1$ the sequence (f_n), defined by $f_n = E^n f$, converges to f in (H^1, p_1) as n tends to ∞. This implies that the subspaces

$$H_o^1 = \{f \in H^1 : f = E^n f \text{ for a } n \geq 1\} = \{f \in L^1 : f = E^n f \text{ for a } n \geq 1\}$$

and $H_o^2 = \{f \in L^2 : f = E^n f \text{ for a } n \geq 1\}$ are dense in (H^1, p_1).

On the other hand H_o^2 is dense in L^2 and in view of the remarks in 4.17 the L^2-norm topology is finer than the p_1-topology, restricted to H_o^2.

Consequently the dual of (H_o^2, p_1) can be identified as the subspace BMO of L^2, defined by

$$\text{BMO} = \{g \in L^2 : \sup \{E(fg) : f \in H_o^2, p_1(f) \leq 1\} < \infty\}.$$

The dual bilinear form is given by $\langle f, g \rangle = E(fg)$ $(f \in H_o^2, g \in \text{BMO})$.

Let us denote by q_1 the norm, dual to p_1 and by q_i^n ($i = 1,2,3$ $n=1,2,\ldots$) the norms on $L_n^\infty = \{g \in L^\infty : E^n g = g\}$, defined as in 4.17 with respect to the finite sequence of σ-algebras $F_1, F_2, \ldots, F_n$.

An element $g \in L^2$ is an element of BMO with $q_1(g) \leq 1$ if and only if $E^n g \in L^\infty$ and $q_1^n(E^n g) \leq 1$ are satisfied for every $n \leq 1$.

We write q_2 for the norm on BMO dual to p_2 and we define a third norm q_3 on BMO, namely

$$q_3(g) = \sup_k \|(E^k(g - E^{k-1}g)^2)^{1/2}\|_\infty$$

According to Jensen's inequality we have for $n \geq 1$

$$[q_3(E^n g)]^2 = [q_3^n(E^n g)]^2 = \sup_k \|E^k[(E^n g - E^{k-1}E^n g)^2]\|_\infty$$

$$\leq [q_3(g)]^2$$

and, since $[E^n(g - E^{k-1}g)]^2$ converges to $(g - E^{k-1}g)^2$ in the L^1-norm and P-almost surely, as n tends to ∞, $E^k[(E^n g - E^{k-1}E^n g)^2]$ converges to $E^k[(g - E^{k-1}g)^2]$ almost surely. This proves that $q_3(g) = \lim_n q_3^n(E^n g)$

36

holds for every $g \in BMO$. In view of Proposition 4.18 q_3 is therefore a norm on BMO, equivalent to q_1 and q_2.

BMO, the dual of (H_o^2, p_1) and hence isometrically isomorphic to the dual of (H_o^1, p_1), is therefore identified as the space $BMO = \{g \in L^2 : q_3(g) < \infty\}$, and we have $<f,g> = E(fg)$ for every $f \in H_o^1$ and $g \in BMO$.

Now every element $f \in H^1$ is the limit of the sequence $(E^n f)$ of elements of H_o^1, the dual bilinear form on the pair (H^1, BMO) can therefore be defined as $<f,g> = \lim_n E(E^n f \, E^n g)$.

Thus we finally get

4.20 Theorem: Let BMO be the space of all martingales $G = (G_n)$ which admit a constant c such that $E[(G_\infty - G_{n-1})^2 | F_n] \le c$ holds for $n = 0, 1, 2, \ldots$ $(G_{-1} = 0)$. Then $q_3(G) = \sup_n \|(E[(G_\infty - G_{n-1})^2 | F_n])^{1/2}\|_\infty$ is a norm on BMO and (BMO, q_3) is equivalent to the strong dual of H^1 under the bilinear form $<F,G> = \lim_n E(F_n G_n)$, $(F = (F_n) \in H^1, G = (G_n) \in BMO)$.

Using Proposition 4.18 we get the following

<u>Corollary:</u> The norms q_i $(i=1,2,3)$ on BMO are related by $q_3 \le \sqrt{5} \, q_1$, $q_1 \le 5 q_2$, $q_2 \le \sqrt{2} \, q_3$. In particular *Fefferman's inequality* holds for $F = (F_n) \in H^1$, $G = (G_n) \in BMO$: $|E(F_n G_n)| \le \sqrt{2} \, p_2(F) \, q_3(G)$ $(n = 1, 2, \ldots)$

As in 4.17 we may imbed (H^1, p_1) into $L_c^1 = L_c^1(P)$, the space of Bochner integrable functions with values in the Banach space c of converging real sequences (β_n) with the norm $\|(\beta_n)\| = \sup_n |\beta_n|$, by means of the imbedding map $\rho_1 : H^1 \to L_c^1$, defined as $\rho_1(F) = (F_n)$.

Similarly (H^1, p_2) can be imbedded into $L_{l_2}^1$, where l_2 is the space of all square summable real sequences, by means of the map ρ_2, defined as $\rho_2(F) = (F_n - F_{n-1})$.

It is not very hard to see that the dual of L_c^1 (resp. $L_{l_2}^1$) is equivalent to $L_{l_1}^\infty$ (resp. $L_{l_2}^\infty$), the space of sequences (h_n) of elements h_n of L^∞, such that $\Sigma_n |h_n|$ (resp. $(\Sigma_n h_n^2)^{1/2}$) are elements of L^∞. The dual bi-linear form is given by $\langle (f_n), (h_n) \rangle = \Sigma_n E(f_n h_n)$.

Consequently (BMO, q_1) can be identified with the quotient $L_{l_1}^\infty / (\rho_1')^{-1}(0)$ and (BMO, q_2) with $L_{l_2}^\infty / (\rho_2')^{-1}(0)$. This gives us

4.21 Theorem:

a) A martingale $G = (G_n)$ is an element of BMO with $q_1(G) \le 1$, if and only if G_∞ has a representation as $G_\infty = \Sigma_k E(h_k | F_k)$, where (h_k) is a sequence of elements of $L^\infty(P)$ satisfying $\Sigma_k |h_k| \le 1$.

b) A martingale $G = (G_n)$ is an element of BMO with $q_2(G) \le 1$, if and only if G_∞ has a representation as $G_\infty = \Sigma_k [E(h_k | F_k) - E(h_k | F_{k-1})]$, where (h_k) is a sequence of elements of $L^\infty(P)$ satisfying $\Sigma_k h_k^2 \le 1$.

§5 CONTINUOUS PARAMETER MARTINGALES

In the previous section we saw that, because of the countability of the parameter set N, to every discrete martingale $M = (M_n)$, given as $L^1(P)$-process, there corresponds an (essentially) unique stochastic process which is a modification of M. The main objective of this section is to establish a similar 1-1 correspondence for continuous parameter martingales. This will be possible because of the 'right continuity' assumption for stochastic bases.

First of all let us fix a stochastic base $(\Omega, F, P, (F_t), R_+)$ and note that we assume that $F_t = F_{t_+}$ holds for all $t \in R_+$.

5.1 Proposition: Every martingale $t \to X_t$ is a right continuous $L^1(P)$-process. A supermartingale $t \to X_t$ is a right continuous $L^1(P)$-process if and only if the real function $t \to E(X_t)$ is right continuous.

Proof: Because for martingales the function $t \to E(X_t)$ is constant, it suffices to prove the second assertion of the proposition. Let (s_k) be a decreasing sequence of real numbers $s_k > t$, converging to $t \in R_+$. Then $E(X_{s_k}) \le E(X_t)$ holds for all k and according to Theorem 4.15 the sequence (X_{s_k}) converges in $L^1(P)$ to an element f.

Now f is F_{t_+} and hence F_t-measurable, we therefore have

$$f = E(f|F_t) = E(\lim_k X_{s_k} | F_t) \le X_t \text{ and}$$

$$E(f) = \lim_k E(X_{s_k}) = E(X_t).$$

Consequently $f = X_t$ holds and the proposition is proved.

Using 4.6 Corollary 1 instead of Theorem 4.15 we get the following.

5.2 Proposition: Every martingale $t \to X_t$ has left hand limits (as $L^1(P)$-process), more precisely there is an (F_t)-adapted martingale $t \to X_{t_-}$, such that $\lim_{s \to t} (s<t) X_s = X_{t_-}$ holds for every $t \in R_+$.

For a supermartingale $t \to X_t$ an analogous statement holds if the set $\{X_t : t \in R_+\}$ is uniformly integrable.

The following theorem establishes regularity properties of the paths of a supermartingale.

5.3 Theorem: Let $X = (X_t)$ be a modification of the supermartingale $t \to X_t$. For a countable dense subset $D \subseteq R_+$ almost every path $X(\omega) : s \to X_s(\omega)$ ($s \in D$) has right and left hand limits for every $t \in R_+$.

Proof: It suffices to prove the theorem for a closed interval $I = [t_0, t_1]$ of R_+ and a dense subset $D \subseteq I$ i.e. we have to prove that except for a set of measure 0

$$\liminf\nolimits_{s \to t} (s \in D, s > t) X_s(\omega) = \limsup\nolimits_{s \to t} (s \in D, s > t) X_s(\omega)$$

and

$$\liminf\nolimits_{s \to t} (s \in D, s < t) X_s(\omega) = \limsup\nolimits_{s \to t} (s \in D, s < t) X_s(\omega)$$

hold for all $t \in I$.

Let D be any countable dense subset of I and observe that by virtue of Theorem 4.10 the inequality $\sup \{x_s : s \in D\} < \infty$ holds P-almost surely. We may therefore assume that all the limits above are finite.

Let us now show the existence of the right hand limits, the proof of the existence of the left hand limits is similar. For any two rationals $a < b$ let $S(a,b)$ be the set of all $\omega \in \Omega$, such that there is a $t \in I$ with

$$\liminf\nolimits_{s \to t} (s \in D, s < t) X_s(\omega) < a \text{ and}$$

$$\limsup\nolimits_{s \to t, \, (s \in D, \, s > t)} X_s(\omega) > b .$$

In the complement of the union $\cup S(a,b)$ over all pairs of rationals $a < b$, the right hand limit exists for every $t \in \overset{o}{I}$.

Let now (D_k) be an increasing sequence of finite subsets $D_k = \{t_1, \ldots, t_k\}$ of D such that $\cup_{k \geq 1} D_k = D$. Fix a pair of rationals $a < b$ and denote by ${}^k U_a^b$ the number of completed upcrossings of the interval $[a,b]$ by the supermartingale $(X_{t_j})_{t_j \in D_k}$ (see 4.5). According to 4.5 Corollary 1 we have the inequality (5.3.1):

$$E({}^k U_a^b) \leq 1/b-a \; E((X_{t_k} - a)^-) .$$

Putting $U_a^b = \lim_k {}^k U_a^b$ we get $S(a,b) \subseteq \{U_a^b = \infty\}$ and, since by (5.3.1) and the Monotone Convergence Theorem $E(U_a^b) < \infty$ holds, $P(S(a,b))$ must be equal to zero for every pair $a < b$. This completes the proof of the theorem.

In view of propositions 5.1 and 5.2 the following corollaries are now evident:

<u>Corollary 1</u>: Let $t \to X_t$ be a martingale resp. a supermartingale such that the real function $t \to E(X_t)$ is right continuous.

There is an (up to indistinguishability) unique right continuous modification $X = (X_t)$ of $t \to X_t$, and this modification has P-almost surely left hand limits.

<u>Corollary 2</u>: Let $t \to X_t$ be a martingale with respect to the family (F_{t-}) $(t \in R_+)$. There is an (up to indistinguishability) unique left continuous modification of $t \to X_t$ and this modification has P-almost surely right hand limits.

Because we are considering equivalence classes of indistinguishable processes rather than individual stochastic processes, Corollary 1 establishes a 1 - 1 correspondence between right continuous martingales (as stochastic processes) and martingales as $L^1(P)$-processes.

By slight modifications of the proof of Theorem 5.3 (observing that the set R_+ is order isomorphic to a half open finite interval $[o, t[$) we get a proof of the assertion a) of the following theorem. The integrability of the limit functions and the assertions b) and c) follow as in the discrete case. In order to be able to state the theorem in a convenient way, we extend our stochastic base:

Put $\bar{R}_+ = R_+ \cup \{\infty\}$ and $F_\infty = \bigvee_{t \in R_+} F_t$, and denote by $(X_t)_{t \in \bar{R}_+}$ a stochastic process with respect to the stochasti base $(\Omega, F, P, (F_t), R_+)$.

5.4 Theorem: Let $t \to X_t$ be a $L^1(P)$-right continuous supermartingale and $X = (X_t)$ ($t \in R_+$) a right continuous modification of $t \to X_t$.

a) If $\sup_{t \in R_+} E(X_t^-) < \infty$ holds, then (X_t) converges almost surely to an integrable random variable X_∞ as t tends to ∞.

b) If $X = (X_t)$ is positive the assertions of a) are satisfied and $(X_t)_{t \in \bar{R}_+}$ is a supermartingale.

c) If $X = (X_t)$ is a uniformly integrable martingale, then (X_t) converges almost surely and in $L^1(P)$-norm to X_∞ as t tends to ∞, and $(X_t)_{t \in \bar{R}_+}$ is a martingale.

The existence of right continuous modifications enables us to define optional sampling for $L^1(P)$-continuous supermartingales.

Let $t \to X_t$ ($t \in R_+$) be a $L^1(P)$-right continuous supermartingale. Define for a stopping time T the random variables X_T, X_{T-} and the stopped process X^T by means of the right continuous modification of $t \to X_t$.

Approximating from above the stopping time T by a sequence of countably valued stopping times T_n, we get as immediate consequence of the discrete Optional Sampling Theorem (Theorem 4.16) the following

5.5 Theorem: Let $t \to X_t$ $(t \in R_+)$ be a $L^1(P)$-right continuous supermartingale (with respect to the extended stochastic base), and let T and S be stopping times such that $S \leq T$ is satisfied.

a) X_T and X_S are elements of $L^1(P)$ and satisfy the inequality
$$E(X_T \mid F_S) \leq X_S.$$

b) If $t \to X_t$ is a martingale, then $X_S = E(X_T \mid F_S) = E(X_\infty \mid F_S)$ holds.

c) The stopped process X^T is again a $L^1(P)$-right continuous supermartingale.

§6 CLASSIFICATION OF STOPPING TIMES

In this section stopping times are studied, in particular predictable ones, i.e. roughly speaking stopping times T, such that T can be approximated from below by a sequence (T_n) of stopping times T_n, strictly smaller than T. (Note that an approximation from above by a sequence of stopping times strictly greater than T on $\{T < \infty\}$ is possible for every stopping time T). The significance of these predictable stopping times T lies in the fact that a closed stochastic interval $[T,S]$ can be written as an inter-section of the half open intervals $]T_n,S]$ (for the definitions see 6.4).

Let $(\Omega, F, P, (F_t), R_+)$ be a stochastic base, i.e. a complete probability space (Ω, F, P) and a right continuous increasing family (F_t) $(t \in R_+)$ of sub σ-algebras of F, such that F_0 contains all P-null sets of F and $\underset{t \in R_+}{\vee} F_t = F$.

A stopping time T with respect to $(F_t)_{t \in R_+}$ was defined as a random variable T on (Ω, F) with values in the extended positive real line $\bar{R}_+ = R_+ \cup \{\infty\}$, satisfying $\{T \leq t\} \in F_t$ for every $t \in R_+$.

The σ-algebra of all events taking place up to time T was defined as

$$F_T = \{A \in F : A \cap \{T \leq t\} \in F_t \text{ for all } t \in R_+\}.$$

We define the σ-algebra F_{T-} of all events taking place strictly before T as the σ-algebra generated by all sets of the form $A \cap \{t < T\}$ where A is an element of F_t and $t \in R_+$.

Analogous to Proposition 3.5 we have

6.1 Proposition: Let S and T be stopping times. Then the following assertions are valid:

a) $F_{T_-} \subseteq F_T$

b) T is F_{T_-} measurable

c) For every $A \in F_S$ $\quad A \cap \{S < T\} \in F_{T_-}$

d) If $T \leq S$ then $F_{T_-} \subseteq F_{S_-}$

Proof: The assertions a), b) and d) are obvious by the definition of F_{T_-}.

To prove c), note that

$$A \cap \{S < T\} = \bigcup_r [A \cap \{S \leq r\} \cap \{r < T\}],$$

where r runs through all positive rationals. Since A is an element of F_S, we have $A \cap \{S \leq r\} \in F_r$ for all r, which proves c).

6.2 Proposition: Let T be a stopping time and A an element of F. Then $A \cap \{T = \infty\} \in F_{T_-}$.

Proof: Let A be an element of F_t. Then $A \cap \{T = \infty\} = \bigcap_{n \geq 1} A \cap \{T \geq t+n\}$, where n runs through all positive integers. Hence $A \cap \{T = \infty\} \in F_{T_-}$ for every $A \in F_{t}$, $t \in R_+$ and, since the family of sets $B \in F$, satisfying $B \cap \{T = \infty\} \in F_{T_-}$ is a sub σ-algebra of $F = \bigvee_{t \in R_+} F_{t}$, this proves the proposition.

Corollary: Let S, T be two stopping times satisfying $S \leq T$. If $S < T$ on $\{0 < T < \infty\}$, then $F_S \subseteq F_{T_-}$.

Proof: For $A \in F_S$ we have

$$A = [A \cap \{T = 0\}] \cup [A \cap \{S < T\}] \cup [A \cap \{T = \infty\}].$$

Now $A \cap \{T = 0\} \in F_0$, $A \cap \{S < T\} \in F_{T_-}$ by Prop. 6.1 and $A \cap \{T = \infty\} \in F_{T_-}$ by Prop. 6.2.

We are now able to characterize F_T resp. F_{T_-} in the following way:

6.3 Theorem: Let (T_n) be a monotone sequence of stopping times with $T = \lim_n T_n$.

a) If (T_n) is decreasing then $F_T = \cap_n F_{T_n}$.

b) If (T_n) is decreasing and $T < T_n$ on $\{0 < T < \infty\}$ for every n, then $F_T = \cap_n F_{T_{n-}}$.

c) If (T_n) is increasing then $F_{T_-} = \vee_n F_{T_{n-}}$.

d) If (T_n) is increasing and $T_n < T$ on $\{0 < T < \infty\}$ for every n, then $F_{T_-} = \vee_n F_{T_n}$.

Proof: a) $\cap_n F_{T_n}$ contains F_T by Proposition 3.5. On the other hand for $A \in \cap_n F_{T_n}$ and for every $t \in R_+$, $A \cap \{T_n < t\}$ belongs to F_t for every n, hence $A \cap \{T < t\}$ belongs to F_t and consequently $A \cap \{T \leq t\}$ is an element of $F_{t_+} = F_t$.

b) is an immediate consequence of a) and the previous corollary.

To prove c), note that F_{T_-} contains $\vee_n F_{T_{n-}}$ by 6.1 d). On the other hand we have for every $t \in R_+$ and $A \in F_t$ the equality $A \cap \{t < T\} = \cup_n A \cap \{t < T_n\}$, which proves that $\vee_n F_{T_{n-}}$ contains a generator of F_{T_-}. d) is an immediate consequence of c) and the previous corollary.

6.4 Definition: A stopping time T is called *predictable* if there exists an increasing sequence (T_n) of stopping times T_n, satisfying a) and b):

a) $T = \lim_n T_n$

b) $T_n < T$ on $\{T > 0\}$ for every n.

Under these conditions (T_n) is called a sequence of stopping times predicting T.

Examples of predictable stopping times are constant stopping times, or more generally, stopping times of the form $T + t$, where T is an arbitrary stopping time.

For two stopping times S and T, satisfying $S \leq T$, the stochastic interval $]S,T]$ is defined as the set

$$]S,T] = \{(t,w) \; R_+ \times \Omega : S(w) < t \leq T(w)\}.$$

Analogously the stochastic intervals $]S,T[$, $[S,T]$ and $[S,T[$ are defined. If $S = T$ we write $[T]$ instead of $[T,T]$. $[T]$ is the graph of the stopping time T.

A stopping time T is called *totally inaccessible*, if $[T] \cap [S]$ is an evanescent set (see 1.2) for every predictable stopping time T.

Finally we call a stopping time *accessible*, if the graph $[T]$ of T is contained up to evanescent sets in the union of the graphs of a sequence of predictable stopping times (T_n).

A totally inaccessible stopping time is P-almost surely strictly greater than zero, every predictable stopping time is accessible, and if a stopping time T is both accessible and totally inaccessible, then $T = \infty$ holds P-almost surely. (We have $\{T < \infty\} = \{w \in \Omega : (t,w) \in [T]$ for a $t \in R_+\}$. Now the graph $[T]$ of T must be an evanescent set and consequently $\{T < \infty\}$ a P-null set).

If T is a stopping time and A an element of F, we call *the restriction of T to A* the random variable T_A, which is equal to T on A, and equal to ∞ on A^c. T_A is a stopping time, if and only if A is an element of F_T. If t is an element of R_+, t_A means the random variable being equal to t on A and equal to ∞ on A^c.

With this notation we are able to state a decomposition theorem of stopping times into an accessible part and a totally inaccessible part.

6.5 Theorem: Let T be a stopping time. Then there exists an essentially unique partition of $\{T < \infty\}$ into two elements A and B of F_{T_-}, such that T_A is accessible and T_B is totally inaccessible.

Proof: Let S be a predictable stopping time and (S_n) a sequence of stopping times predicting S. Then $\{S \leq T\} = \bigcap_n \{S_n < T\} \cup \{S = T = 0\}$.

Hence $\{S \leq T\} = \{S = T\} = \{S \leq T\} \smallsetminus \{S < T\}$ are elements of F_{T_-} by Prop. 6.4 c). Denote by X_A a representant of the essential supremum of the family $\{X_{\{S=T\} \cap \{T<\infty\}}: S$ is a predictable stopping time$\}$. Because F_T is a complete σ-algebra, the set A is an element of F_T. Consequently T_A is an accessible and $T_B = T_{\{T<\infty\}\smallsetminus A}$ is a totally inaccessible stopping time. The uniqueness of the decomposition is an immediate consequence of the definitions (6.4) and the following

6.6 Proposition: Let T be a stopping time and $A \in F_T$. If T is accessible (resp. totally inaccessible), then T_A is accessible (resp. totally inaccessible).

The proof of this proposition is obvious from the fact that the graph of the restriction T_A of T is contained in the graph of T.

The following theorem shows that the family of predictable (resp. accessible) stopping times forms a sublattice of the ordered lattice of stopping times, which is closed under suprema of increasing sequences.

6.7 Theorem: a) Let S and T be two stopping times. Then $S \vee T$ and $S \wedge T$ are predictable (resp. accessible) if both S and T are predictable (resp. accessible).

b) Let (T_n) be an increasing sequence of predictable (resp. accessible) stopping times. Then $T = \lim_n T_n$ is predictable (resp. accessible).

c) Let (T_n) be a decreasing sequence of predictable (resp. accessible) stopping times with $T = \lim_n T_n$. If for P-almost every $\omega \in \Omega$ $T(\omega) = T_n(\omega)$ for a suitable n, then T is predictable (resp. accessible).

<u>Proof:</u>

a) is obvious from the definitions 6.4.

b) put $A = \{\omega \in \Omega : T_n(\omega) < T(\omega) \text{ for all } n\}$. Then T_A is predicted by the sequence (R_n) of stopping times $R_n = T_A \wedge T_n \wedge n$, i.e. T_A is predictable and hence accessible. Now $T = T_A \wedge T_B$ is accessible by a).

In order to prove the predictable case, note that for every n there exists a sequence of stopping times $(T_{n,p})_p$, predicting T_n. Then the sequence (T_k) of stopping times $T_k = \sup_{p \leq k, n \leq k} T_{n,p}$ is a sequence of stopping times predicting T.

c) is obvious in the accessible case.

For the predictable case let $(T_{n,p})_p$ be a sequence of stopping times predicting T_n for every n, such that

$$P(\{\omega \in \Omega : d(T_{n,p}(\omega) - T_n(\omega)) > 2^{-p}\}) \leq 2^{-(n+p)}$$

for every p is valid, where $d(t,s) = \arctan(|t-s|)$ and $d(t,\infty) = d(\infty,t) = \frac{\pi}{2}$ for $t,s \in \mathbb{R}_+$.

Put $S_p = \inf_n S_{n,p}$. Then $S_p < T$ on $\{T > 0\}$ and

$$P(\{\omega \in \Omega : d(S_p(\omega) - T(\omega)) > 2^{-p}\})$$

$$\leq \sum_n P\{\{\omega \in \Omega : d(T_{n,p}(\omega) - T_n(\omega)) > 2^{-p}\}) \leq 2^{-p}$$

This proves that (S_p) is a sequence of stopping times predicting T.

Remark: Note that every stopping time T can be represented as the infimum of a decreasing sequence (T_n) of predictable stopping times. (Take for example the sequence $(T_n) = (T + \frac{1}{n})$.)

As an immediate consequence of the previous theorem, we have that the family of elements $A \in F_T$, such that the restricted stopping time T_A is predictable, is stable under countable unions and intersections. This result allows us to prove a proposition for predictable stopping times analogous to 6.6.

6.8 Proposition: Let T be a predictable stopping time, A and element of F_T. Then T_A is predictable if and only if A is an element of F_{T_-}.

Proof: Suppose T_A is predictable and let (S_n) be a sequence of stopping times predicting T_A.

Then $A = \{T_A \leq T\} \smallsetminus (A^c \cap \{T = \infty\})$

$= \cap_n \{S_n < T\} \smallsetminus (A^c \cap \{T = \infty\})$.

Now $\{S_n < T\} \in F_{T_-}$ by 6.1 c) and $(A^c \cap \{T = \infty\}) \in F_{T_-}$ by Proposition 6.2. This proves $A \in F_{T_-}$.

The prove that for every $A \in F_{T_-}$ T_A is predictable consider the family of all elements $A \in F_T$ such that T_A and T_{A^c} are predictable. Since this family forms a σ-algebra by the remark above, it suffices to show that it contains a generator of F_{T_-}. Let (T_n) be a sequence of stopping times predicting T. For every $A \in F_{T_m}$, the family of stopping times (S_n), where $S_n = T_{n+m, A}{\wedge}n$, predicts T_A, i.e. T_A is predictable for every $A \in \cup_n F_{T_n}$. Now $F_{T_-} = \vee_n F_{T_n}$ and this proves the proposition.

§7 BASIC SUB σ-ALGEBRAS OF $\Sigma = \mathcal{B} \otimes F$ AND PROJECTIONS OF STOCHASTIC PROCESSES.

For this section a stochastic base $(\Omega, F, P, (F_t), R_+)$ is fixed. Denote by $\mathcal{B}$ the Borel σ-algebra on R_+ and by Σ the product σ-algebra $\mathcal{B} \otimes F$ on $R_+ \times \Omega$.

In §1 a measurable stochastic process was defined as random variable on $(R_+ \times \Omega, \Sigma)$. In §3 the sub σ-algebra Σ_o of Σ of progressive sets was introduced, and it was shown, that adapted right (resp. left) continuous processes are progressively measurable.

In this section we study the σ-algebras Σ_w (resp. Σ_p) generated by right (resp. left) continuous processes. The σ-algebra Σ_p will be of particular importance for the following, because stochastic integration will be defined with respect to a large class of stochastic processes for Σ_p-measurable functions.

The main tools for this investigation are capacity theorems. (For a treatment of capacity theory see [4] chap. I).

In general the projection of a measurable set in a product space onto a component need not be measurable. The next theorem shows that this result holds in our special situation. Moreover, the section theorem 7.2 shows that the projection of a measurable set B of the product space can be written as the projection of the graph of a measurable mapping from the second component to the first one. Both theorems are quoted without proof from [4].

<u>7.1 Theorem:</u> ([4] chap. I T32)

Let (Ω, F, P) be a complete probability space. For every element B of Σ, the projection $\pi(B)$ of B, where π denotes the projection from $R_+ \times \Omega$ onto Ω, is an element of F.

7.2 Theorem: ([4] chap. I T37)

Let (Ω, F, P) be a complete probability space and B an element of Σ. There is a random variable Z with values in the extended positive real line $\bar{R}_+$, such that the following assertions are valid:

(a) The graph $Z = \{(t,\omega) \in R_+ \times \Omega : Z(\omega) = t\}$ of Z is contained in B.

(b) The sets $\{\omega : Z(\omega) < \infty\}$ and $\pi(B)$ coincide.

Theorem 7.2 is known as *section theorem*.

Let now A be a subset of $R_+ \times \Omega$. The function $D(A)$, defined as

$$D(A)(\omega) = \inf \{t \in R_+ : (t,\omega) \in A\}$$

is called *debut of* A. (As usual, the infimum of the empty set is equal to ∞).

The following proposition is a consequence of Theorem 7.1:

7.3 Proposition: For an element $A \in \Sigma$ the debut $D(A)$ of A is an F-measurable function.

Proof: We have $\{D(A) < t\} = \pi(A \cap [0,t[) $ and $\pi(A \cap [0,t])$ is an element of F according to Theorem 7.1.

Corollary 1: For a progressive set $A \in \Sigma_0$ the debut $D(A)$ of A is a stopping time.

Proof: $A \in \Sigma_0$ implies $A \cap [0,t[\in B_t \otimes F_t$, where B_t is the Borel σ-algebra on $[0,t]$. By the proposition above, $\{D(A) < t\} = \pi(A \cap [0,t[)$ is an element of F_t and, since the family (F_t) of σ-algebras is right continuous, $D(A)$ is a stopping time.

<u>Corollary 2</u>: Let $X = (X_t)$ be a progressively measurable stochastic process. For every Borel subset B of R the random variable

$$Z(\omega) = \inf\{t > o : X_t(\omega) \in B\}$$

is a stopping time.

The stopping time Z defined in the corollary above is called the *first hitting time of* B.

<u>Proof</u>: Put $A = \{(t,\omega) : X_t(\omega) \in B\}$. A is a progressive set and so is the stochastic interval $]o,\infty[$ according to Theorem 3.3. Hence the debut $D(A \cap]o,\infty[)$ of the set $A \cap]o,\infty[$, which is equal to Z, is a stopping time.

Let us now define the basic sub σ-algebras of Σ :

<u>7.4 Definition</u>: The σ-algebras Σ_w (resp. Σ_a, Σ_p), generated by all stochastic intervals of the form $[S,T[$, where $S \leq T$ and S,T are arbitrary (resp. accessible, predictable) stopping times, are called the σ-algebras of well measurable (resp. accessible, predictable) sets.

We have $\Sigma_p \subseteq \Sigma_a \subseteq \Sigma_w$ and, since the processes $X_{[S,T[}$ are right continuous and adapted and hence progressively measurable by 3.3 Theorem, we have $\Sigma_w \subseteq \Sigma_o$.

Since every stopping time T is the limit of a sequence (T_n) of predictable stopping times $T_n > T$ on $\{T < \infty\}$, Σ_w (resp. Σ_a, Σ_p) is also generated by the family of stochastic intervals of the form $[S,T]$, where $S \leq T$ and S,T are arbitrary (resp. accessible, predictable) stopping times.

Moreover Σ_p is generated by stochastic intervals of the form $[O_A] = \{o\} \times A$, where $A \in F_o$; and $]S,T]$, where S,T are arbitrary stopping times satisfying $S \leq T$.

It is not hard to see, that every stopping time T is the limit of a decreasing sequence (T_n) of stopping times T_n, taking only finitely many values and being strictly greater than T on $\{T < \infty\}$. Consequently Σ_p is generated by stochastic intervals of the form $[0_A]$ $(A \in F_o)$ and $]s_B, t_B] =]s,t] \times B$ $(B \in F_s)$.

Denote by A^i $(i = 1,2,3)$ the Boolean algebra generated by all stochastic intervals $[S,T[$, where S and T are arbitrary stopping times in case $i = 1$, accessible stopping times in case $i = 2$, and predictable stopping times in case $i = 3$, such that $S \leq T$.

The following lemma will be needed to prove a theorem analogous to the section theorem 7.2, where Z is replaced by a stopping time. This theorem will enable us to show, that certain processes are determined by their values at stopping times.

7.5 Lemma: Let (A_n) be a decreasing sequence of elements of A^i $(i=1,2,3)$ And $A = \bigcap_n A_n$. The debut $D(A)$ of A is a stopping time for $i = 1$, an accessible stopping time for $i = 2$ and a predictable stopping time for $i = 3$, and the graph $[D(A)]$ of $D(A)$ is contained in A.

Proof: It is clear that the graph $[D(A)]$ of the debut of A is contained in A, because A^i is generated by stochastic intervals closed to the left for every $i = 1,2,3$. Let us prove the Lemma for $i = 3$, the proofs for other cases are similar. For a stochastic interval $[S,T[$ the debut $D([S,T[)$ is equal to the restricted stopping time $S_{\{S<T\}}$. Now $\{S < T\}$ is an element of F_{T-} by 6.2 c) and, since S is a predictable stopping time, $D([S,T[)$ is predictable by Proposition 6.8. The elements of A^3 are finite unions of stochastic intervals $[S,T[$ with predictable S, consequently the debuts of elements of A^3 are infimums of finite families of predictable stopping times, by 6.7 a) these debuts are thus predictable.

Consider now an element $A \in \Sigma_p$ of the form $A = \bigcap_n A_n$, where (A_n) is a decreasing sequence of elements of A^3. Denote by S the family of all predictable stopping times, smaller than $D(A)$, the debut of A. S is nonempty and stable under countable suprema by Theorem 6.7. Every representant T of the essential supremum of S therefore is a predictable stopping time and T moreover almost surely equal to $D(A)$: replacing if necessary A_n by $B_n = A_n \cap [T, \infty[$, we have $A = \bigcap_n B_n$ and $D(B_n) \geq T$ for every n. Now $D(B_n)$ is an element of S for every n, hence $D(B_n)$ is almost surely equal to T for every n. Since the graph of $D(B_n)$ is contained in B_n, we have $\sup_n D(B_n) = D(A)$ and finally $T = \sup_n D(B_n) = D(A)$ P-almost surely.

We use Lemma 7.5 to prove the following section theorem:

7.6 Theorem: Let A be a well measurable (resp. accessible, predictable) set. For every $\varepsilon > 0$ there exists a stopping time T (resp. accessible stopping time T, predictable stopping time T) such that the following assertions hold:

(a) $[T] \subseteq A$.

(b) $P(\pi(A)) \leq P(\{T < \infty\}) + \varepsilon$ where π is the projection of $\mathbb{R}_+ \times \Omega$ to Ω.

Proof: There exists a positive random variable Z, such that the graph of Z is contained in A and the sets $\pi(A)$ and $\{Z < \infty\}$ coincide by Theorem 7.2. Define the measure μ on $(\mathbb{R}_+ \times \Omega, \Sigma)$ by

$$\mu(E) = P(\{Z < \infty\} \cap \{\omega : (Z(\omega), \omega) \in E\}).$$

In particular we have $\mu(A) = P(\pi(A))$.

On the other hand since the algebras A^i ($i = 1, 2, 3$) are generators of the respective σ-algebras, there exists a decreasing sequence (A_n) of elements of A^i $i = 1$ (resp. $i = 2$, $i = 3$), such that $\cap_n A_n = B \subseteq A$ and $\mu(A) \leq \mu(B) + \varepsilon$. We therefore have for $T = D(B)$ the debut of B:

$$P(\pi(A)) = \mu(A) \leq \mu(B) + \varepsilon \leq P(\{T < \infty\}) + \varepsilon.$$

The following corollary provides an easy way to check the indistinguishability of certain stochastic processes:

<u>Corollary 1</u>: Let X and Y be two well measurable (resp. accessible, predictable) stochastic processes. X and Y are indistinguishable if and only if $X_T = Y_T$ is valid almost surely for every finite and well measurable (resp. accessible, predictable) stopping time.

<u>Proof</u>: The proof is immediate from Theorem 7.6 taking into account the fact that the set $\{(t,w) : X_t(w) \neq Y_t(w)\}$ is an element of Σ_w (resp. Σ_a, Σ_p).

The following corollary characterizes stopping times by means of their graphs:

<u>Corollary 2</u>: A positive random variable T is a stopping time (resp. accessible, predictable stopping time) if and only if the graph $[T]$ of T is a well measurable (resp. accessible, predictable) set.

<u>Proof</u>: The necessity of the conditions is a consequence of the definitions 7.3.

For the sufficiency in the well measurable case, note that the debut $D([T])$ of the graph of T coincides with T, and is a stopping time by 7.3 Corollary 1. For the other cases there exists a sequence (T_n) of accessible (resp. predictable) stopping times T_n, such that $T_n \subseteq T$ and $P(\{T < \infty\}) \leq P(\{T_n < \infty\}) + 2^{-n}$ for every n by Theorem 7.6. Replacing T_n by $\inf_{k \leq n} T_k$, we may assume that (T_n) satisfies the conditions of 6.7 c), which proves that T is accessible (resp. predictable).

Corollary 3: If an accessible (resp. predictable) set A contains the graph of its debut $D(A)$, then $D(A)$ is an accessible (resp. predictable) stopping time.

Proof: $]D(A),\infty]$ is predictable, hence $[D(A)] = A -]D(A),\infty[$ is accessible (resp. predictable), and by the previous corollary $D(A)$ is accessible (resp. predictable).

Corollary 4: Every well measurable (resp. accessible, predictable) set A, which is contained in a countable union $\cup_n [S_n]$ of graphs of stopping times S_n, is equal to a disjoint union of graphs of a sequence (T_n) of arbitrary (resp. accessible, predictable) stopping times.

Proof: We may assume that the graphs $[S_n]$ are disjoint, replacing if necessary S_n by the restricted stopping time $D(B_n)$, where $B_n = (A - \cup_{k \leq n} [S_k]) \cap [S_n]$. Denote U_n resp. V_n the accessible resp. totally inaccessible part of S_n (see 6.5 Theorem). Then

$$\cup_n [V_n] = A - \cup_n [U_n]$$

is accessible if A is accessible by Corollary 1. Now $\cup_n [V_n]$ cannot contain a graph of an accessible stopping time other than $T = \infty$ almost surely. $\cup_n [V_n]$ is therefore evanescent by Theorem 7.6, which proves the accessible case.

For a predictable set A this gives us that A is contained in the union of graphs of a countable family of predictable stopping times. Using the same technique as at the beginning of the proof, we can construct a sequence of predictable stopping times with disjoint graphs, the union of which is A.

The following proposition hints at the importance of the sets, studied in 7.6 Corollary 4:

7.7 Proposition: For every well measurable process X there exists a predictable process Y, such that the set $\{X \neq Y\}$ is contained in the union of the graphs of a sequence of stopping times.

Proof: Denote by H the family of all well measurable processes, satisfying the assertion of 7.7. Then H is a vector space which is closed under (pointwise) convergence of sequences of elements of H and contains the constants, i.e. H is a monotone class. According to 7.4, H contains characteristic functions of stochastic intervals $[S,T[$, where S and T are stopping times. The family of these characteristic functions forms a uniformly bounded subset of H which is closed under multiplication, hence H contains all well measurable process by the monotone class theorem.

Another consequence of the monotone class theorem and 7.6 Corollary 4 is the following characterization of predictable processes:

7.8 Theorem: An accessible process X is predictable, if and only if for every predictable stopping time T the random variable $X_T \chi_{\{T<\infty\}}$ is F_{T-} measurable.

Proof: The family of stochastic processes satisfying "$X_T \chi_{\{T<\infty\}}$ is F_{T-} measurable" form a monotone class. In order to prove the necessity, it suffices therefore, that the characteristic functions of intervals of the form $[0_A]$ ($A \in F_0$) and $]U,V]$ (U,V stopping times satisfying $U \leq V$) satisfy the assertion above. That is obvious for the stochastic intervals $[0_A]$, and for $X = \chi_{]U,V]}$ we have:

$$X_T \chi_{\{T<\infty\}} = \chi_{(\{U < T\} - \{V < T\}) \cap \{T < \infty\}}$$

which is F_{T-} measurable by 6.1c) .

In order to prove that the condition is sufficient, notice that there exists a predictable process Y, such that $\{Y \neq X\}$ is contained in the union of the graphs of a sequence of accessible stopping times by Proposition 7.7 and Corollary 4. Therefore there exists a sequence (S_n) of predictable stopping times S_n with disjoint graphs, satisfying

$$\{Y \neq X\} \subseteq \bigcup_n [S_n] .$$

Since $X_{S_n} X_{\{S_n < \infty\}}$ is F_{S_n-} measurable by hypothesis and S_n is a predictable stopping time, the process $X_{S_n} X_{[S_n]}$ is predictable by Proposition 6.8 and 7.6 Corollary 2. By 7.6 Corollary 2 $A = \bigcup_n [S_n]^c$ is a predictable set and hence

$$X = Y X_A + \sum_n X_{S_n} X_{[S_n]}$$

is predictable.

We are now able to prove a useful variante of 7.6 Corollary 1:

<u>7.9 Theorem</u>: Let X and Y be two finite well measurable (resp. predictable, accessible) stochastic processes. The following assertions are equivalent:

(a) $X \geq Y$ up to indistinguishability.

(b) If for an arbitrary (resp. predictable, accessible) stopping time T $X_T X_{\{T < \infty\}}$ and $Y_T X_{\{T < \infty\}}$ are integrable, then $E(X_T X_{\{T < \infty\}}) \geq E(Y_T X_{\{T < \infty\}})$.

<u>Proof</u>: (a) ⇒ (b) is obvious, let us prove the implication (b) ⇒ (a) . We shall prove the predictable case. Suppose $\{X < Y\}$ is not evanescent. By Theorem 7.6 there exists a predictable stopping time T, such that $X_T X_{\{T < \infty\}} < Y_T X_{\{T < \infty\}}$ and $P\{T < \infty\} > 0$ hold. Since X and Y are

finite, there exists a constant k, such that the probability of the set

$$B = \{T < \infty\} \cap \{|X_T| \leq k\} \cap \{|Y_T| \leq k\}$$

is nonzero. By Theorem 7.8 B is a F_{T-} measurable set, the restricted stopping time T_B is therefore predictable by Proposition 6.8, and

$$E(X_{T_B} \chi_{\{T_B < \infty\}}) < E(Y_{T_B} \chi_{\{T_B < \infty\}})$$

holds, which contradicts (b).

The following two theorems show, that Σ_p is the σ-field generated by the family of left continuous adapted processes, and that Σ_w is the σ-field generated by the family of right continuous adapted processes.

7.10 Theorem: The σ-field Σ_p of predictable sets coincides with the σ-field generated by the family of left continuous adapted stochastic processes.

Proof: Σ_p is generated by processes of the form

$$\chi_{[0_A]} \, (A \in F_o) \text{ and } \chi_{]S,T]} \, (S \leq T \text{ stopping times}).$$

Since both processes are left continuous, it suffices to prove, that every left continuous process X is predictable. Put

$$X^n = X_o \chi_{[o]} + \sum_{k \geq 1} X_{k/n} \chi_{]k/n, \, (k+1)/n]}.$$

Then X^n is predictable for every positive integer n, and, since X is left continuous,

$$\lim_n X^n(t,w) = X(t,w)$$

holds for all $(t,w) \in R_+ \times \Omega$, which shows that X is predictable.

Corollary: Σ_p is generated by the family of all continuous adapted processes.

Proof: The process $X_{[0_A, \infty[}$ is continuous and adapted for $A \in F_0$ and a stochastic interval of the form $]S, \infty]$ coincides with the set $\{X > 0\}$ of the continuous process (X_t), defined by $X_t(\omega) = t - S(\omega) \wedge t$.

7.11 Theorem: Σ_w is generated by the family of all right continuous processes which are adapted and have left hand limits.

Proof: As in the proof of Theorem 7.10, it suffices to prove that every right continuous adapted process with left hand limits is well measurable. For a given process $X = (X_t)$ and every positive integer k, define an increasing sequence $(T_n^k)_n$ by induction:

$$T_1^k = 0$$

$$T_{n+1}^k(\omega) = \inf\{t > T_n^k(\omega) : |X_t(\omega) - X_{T_n^k}(\omega)| \geq 1/k\}$$

$$(T_{n+1}^k(\omega) = \infty \text{ if this set is empty or if } T_n^k(\omega) = \infty)$$

T_{n+1}^k is a stopping time as debut of the progressive set

$$]T_n^k, \infty] \cap \{|X - X_{T_n^k} X_{]T_n^k, \infty]}| \geq 1/k\} \ .$$

Since X is right continuous we have $|X_{T_{n+1}^k} - X_{T_n^k}| > 1/k$ on $\{T_{n+1}^k < \infty\}$, so that the existence of left hand limits implies $\sup_n T_n^k = \infty$ for every k. Now put $X^k = \Sigma_n X_{T_n^k} X_{[T_n^k, T_{n+1}^k[}$ for every k. In view of the right continuity of X, the sequence (X^k) of well measurable processes X^k converges pointwise to the process X, which is therefore well measurable.

Remark: By using a transfinite sequence $(T_\alpha^k)_\alpha$ of stopping times in the previous proof, it can be shown that every right continuous adapted process is well measurable.

We saw in Proposition 7.6 that every well measurable process differs from a suitable predictable process only on the union of the graphs of a sequence of stopping times. This fact was used in Theorem 7.8 to characterize the accessible processes which are predictable. We now study, for a right continuous process $X = (X_t)$ with left hand limits, the difference between X and the process $Y = (X_{t-})$, and this will provide us with a characterisation of the accessible right continuous adapted processes with left hand limits.

7.12 Definition: Let $X = (X_t)$ be an adapted right continuous process with left hand limits.

X *charges* a stopping time T, if $P(\{X_T \neq X_{T-}\} \cap \{T < \infty\}) > 0$, and X *has a jump* in T if X charges T and the sets $\{T < \infty\}$ and $\{X_T \neq X_{T-}\}$ are almost surely equal.

A sequence (T_n) of stopping times *exhausts the jumps of* X, if the following three conditions are satisfied:

(a) X has a jump in T_n for every n.

(b) The graphs T_n are disjoint.

(c) If the graph of a stopping time T is disjoint from $\cup_n [T_n]$, then X does not charge T.

7.13 Theorem: Let $X = (X_t)$ be a right continuous adapted process with left hand limits. There exists a sequence (T_n) of stopping times T_n exhausting the jumps of X. If X is accessible the stopping times T_n are accessible, if X is predictable the stopping times T_n can be taken predictable.

Proof: Write $Y = (Y_t)$ for the process $Y_t = X_{t-}$ and $A = \{X \neq Y\}$. By

7.6 Corollary 4 it suffices to show that A is contained in the graphs of a sequence of stopping times.

For every positive integer n define a sequence of stopping times $(T_{n,k})_{k \geq 0}$ by induction:

$$T_{n,0} = 0$$

$$T_{n,k+1}(\omega) = \inf\{t > T_{n,k}(\omega) : |X_t(\omega) - Y_t(\omega)| > 1/n\},$$

where $T_{n,k+1}(\omega) = \infty$ if this set is empty or if $T_{n,k}(\omega) = \infty$. As in the proof of Theorem 7.11 the existence of left hand limits of the process X yields

$$\sup_k T_{n,k}(\omega) = \infty \quad \text{for every } \omega \in \Omega,$$

and from the right continuity of X we conclude that A is contained in the union of the graphs $\bigcup_{n,k} [T_{n,k}]$.

Corollary 1: Let $X = (X_t)$ be a right continuous adapted process with left hand limits. X is accessible if and only if X does not charge any totally inaccessible stopping time.

Proof: Denote by (T_n) a sequence of stopping times exhausting the jumps of X. If X is accessible these stopping times are accessible too, hence the condition is necessary.

In order to prove the sufficiency, write Y for the process $Y_t = X_{t-}$, and observe that

$$X = X_A Y + \Sigma(X_{T_n} - X_{T_n-})X_{[T_n]} \quad \text{holds, where}$$

$$A = \bigcap_n [T_n]^c.$$

Since X doesn't charge any totally inaccessible stopping time, the stopping times T_n are accessible for every n, hence the set A and the processes $(X_{T_n} - X_{T_{n-}})X_{[T_n]}$ are accessible. Since the adapted left continuous process Y is accessible, this proves that X is accessible.

An immediate consequence of 7.13 Corollary 1 and Theorem 7.7 is the following characterization of adapted processes $V = (V_t)$ of integrable variation in terms of the measures μ_V generated by V on Σ according to $\mu_V(]s,t] \times F) = E(\int x_{]s,t] \times F}(u) \, dV_u)$ (see 2.3).

7.14 Proposition: Let $V \in SV$ be an adapted process of integrable variation and denote by μ_V the σ-additive real measure generated by V on Σ.

(a) V is accessible if and only if the graphs of all totally inaccessible stopping times are μ_V-null sets.

(b) If V is accessible, then V is predictable if and only if $\int (f - E(f|F_{T-}))X_{[T]} \, d\mu_V = 0$ holds for every $f \in L^\infty(P)$ and every predictable stopping time T.

We conclude this section by establishing the existence of order continuous projections between the spaces of all bounded measurable processes and the spaces of all bounded well measurable (resp. accessible, predictable) stochastic processes.

Denote by $B^\infty(\Sigma)$ (resp. $B^\infty(\Sigma_w)$, $B^\infty(\Sigma_a)$, $B^\infty(\Sigma_p)$) the Banach lattice of (equivalence classes) of (essentially) bounded stochastic processes which are measurable with respect to Σ (resp. Σ_w, Σ_a, Σ_p). (Equivalence of two processes X and Y means that X and Y are indistinguishable, but keep in mind that our notation doesn't distinguish between stochastic processes and the associated equivalence class). The norm on the spaces B^∞ is the essential supremum norm. Sequential order continuity of a positive linear operator on B^∞ means that the operator maps every decreasing sequence (X^n)

65

of processes X^n with $\inf_n X^n = 0$ to a sequence (Y^n) with $\inf_n Y^n = 0$.

<u>7.15 Theorem:</u> There exists a unique linear projection π of $B^\infty(\Sigma)$ onto $B^\infty(\Sigma_w)$ (resp. $B^\infty(\Sigma_a)$, $B^\infty(\Sigma_p)$) satisfying (7.15.1):

$$E(X_T \chi_{\{T < \infty\}}) = E((\pi X)_T \chi_{\{T < \infty\}})$$

for every stopping time (resp. accessible stopping time, predictable stopping time) T.

This projection has the following properties (in all cases):

(a) π is positive and sequentially order continuous.

(b) π has norm 1.

(c) If X and Y are elements of $B^\infty(\Sigma)$, then

$$[(\pi X)Y] = (\pi X)(\pi Y) \text{ holds.}$$

We denote the projections onto $B^\infty(\Sigma_w)$ (resp. $B^\infty(\Sigma_a)$, $B^\infty(\Sigma_p)$) by π_w (resp. π_a, π_p). π_w is called the *well measurable projection*, π_a the *accessible* and π_p the *predictable projection*.

<u>Proof:</u> If a map π satisfying 7.15.1 exists, the assertions (a) and (b) and the assertion that π is a linear projection onto the respective spaces, uniquely determined by (7.15.1), is an immediate consequence of Theorem 7.9.

If we denote by H_w (resp. H_a, H_p) the families of all measurable processes admitting a projection that satisfies (7.15.1), then it is easy to see that H_w, H_a and H_p are monotone classes. Consequently, by the monotone class theorem, it suffices to prove the existence of π only for the elements of a uniformly bounded family of measurable processes, which is stable under multiplication and generates the σ-field Σ on $R_+ \times \Omega$.

(i) well measurable case:

Let X be a process of the form $X = Z\chi_{[r,s]}$, where Z is a bounded

F-measurable function, and r,s are two positive real numbers satisfying $r \leq s$.

Denote by Y a right continuous modification of the martingale
$$t \to E(Z|F_t) .$$
Then $\pi_w X = Y X_{[r,s]}$ satisfies (7.15.1) by Doobs optional sampling theorem.

(ii) predictable case: Let $X = Z X_{[r,s]}$ be given as in (i). The martingale $E(Z|F_t)$ has a version which is right continuous and has left hand limits. Denote by Y_- the left continuous process $Y_- = (Y_{t_-})$. For a predictable stopping time T we have by Doobs optional sampling theorem, the convergence theorem 4.6 Corollary 1 and Theorem 6.3 (d):
$$Y_{T_-} = E(Z|F_{T_-}) .$$
This proves that $\pi_p X = Y_- X_{[r,s]}$ satisfies (7.15.1).

(iii) The accessible case: If the accessible projection π_a exists, it satisfies $\pi_a X = \pi_a \pi_w X$ in view of (7.15.1). We may therefore restrict ourselves to proving the existence of π_a for every well measurable process, and, taking into account the monotone class theorem, it suffices to consider processes of the form $X_{[S,\infty[}$, where S is a stopping time.

Now let S_A (resp. S_I) be the accessible be the accessible (resp. totally inaccessible) part of S (Theorem 6.5), and put $B = [S_A, \infty[\cup]S_I, \infty[$. B is accessible as the union of an accessible and a predictable set, and
$$B \cap [T] = [S, \infty[\cap [T]$$
holds for every accessible stopping time, which proves that X_B is the accessible projection $\pi_a X$ of $X = X_{[S,\infty[}$

The assertion (c) is a consequence of the following proposition:

7.16 Proposition: For a bounded measurable process X the following assertions are valid:

(a) $({}^{\pi}{}_{w}X)_T X_{\{T<\infty\}} = E(X_T X_{\{T<\infty\}} | F_T)$ for every stopping time.

(b) $({}^{\pi}{}_{a}X)_T X_{\{T<\infty\}} = E(X_T X_{\{T<\infty\}} | F_T)$ for every accessible stopping time T.

(c) $({}^{\pi}{}_{p}X)_T X_{\{T<\infty\}} = E(X_T X_{\{T<\infty\}} | F_{T_-})$ for every predictable stopping time T.

<u>Proof</u>: Let A be an element of F_T in the cases (a) and (b), and an element of F_{T_-} in the case (c). If T is an arbitrary (resp. accessible, predictable) stopping time, its restriction T_A to A has the respective properties.

From (7.15.1) we have

$$E(X_{T_A} X_{\{T_A<\infty\}}) = E((\pi X)_{T_A} X_{\{T_A<\infty\}})$$

or equivalently

$$E(X_T X_A X_{\{T<\infty\}}) = E((\pi X)_T X_A X_{\{T<\infty\}})$$

in all cases. Since $({}^{\pi}{}_{w}X)_T$ and $({}^{\pi}{}_{a}X)_T$ are F_T-measurable random variables and, since $({}^{\pi}{}_{p}X)_T$ is F_{T_-}-measurable for predictable stopping times T, the last equality proves the proposition.

§8 PROCESSES OF BOUNDED VARIATION AND PROJECTIONS

The projection theorem (Theorem 7.15) allows a unique extension of real stochastic measures on one of the σ-algebras Σ_w, Σ_a, or Σ_p to the σ-algebra Σ.

Let us denote by $sca(\Sigma)$ (resp. $sca(\Sigma_w)$, $sca(\Sigma_a)$, $sca(\Sigma_p)$) the Banach spaces of real stochastic measures on Σ (resp. $\Sigma_w, \Sigma_a, \Sigma_p$) (see §2). As in §7 $B^\infty(\Sigma)$ (resp. $B^\infty(\Sigma_w)$, $B^\infty(\Sigma_a)$, $B^\infty(\Sigma_p)$) are the vector spaces of (equivalence classes of indistinguishable) bounded stochastic processes, measurable with respect to the respective σ-algebras. These spaces are Banach spaces (Banach lattices) under the essential supremum norm.

It is clear that for every element $\mu \in sca(\Sigma_x)$ ($x = w, a, p$) the linear form $Z \to \int \pi_x Z d\mu$ defines a measure on Σ and by virtue of 7.15 (a) this measure is σ-additive and hence a stochastic measure.

This yields the following diagram for $x = w, a, p$:

$\pi_x : B^\infty(\Sigma) \to B^\infty(\Sigma_x)$

$\pi'_x : sca(\Sigma_x) \to sca(\Sigma)$:

where π'_x is the restriction of the adjoint of π_x to $sca(\Sigma_x)$.

The extension $\bar\mu$ to Σ of a measure $\mu \in sca(\Sigma_x)$ is defined as $\bar\mu = \pi'_x \mu$. It is clear that a measure $\bar\mu \in sca(\Sigma)$ is in the range of π'_x i.e. is an extension of a measure $\mu \in sca(\Sigma_x)$, if and only if $\bar\mu$ commutes with the projection π_x in the sense that

$\int Z d\bar\mu = \int \pi_x Z \, d\bar\mu$ holds for every process $Z = (Z_t) \in B^\infty(\Sigma)$.

Furthermore, since the ranges of the projections π_x ($x = w, a, p$) are the spaces $B^\infty(\Sigma_x)$, the extension $\bar\mu$ of a measure $\mu \in sca(\Sigma_x)$ is positive if and only if μ is a positive measure.

We summarize in the following

<u>8.1 Theorem</u>: Every stochastic measure μ on the σ-algebra Σ_w (resp. Σ_a, Σ_p)

has a unique extension $\bar{\mu}$ to a stochastic measure on Σ, which commutes with the projection π_w (resp. π_a, π_p).

The mapping $\mu \to \bar{\mu}$ is an isometric isomorphism onto a Banach sublattice of $sca(\Sigma)$.

According to Theorem 2.6 we have a one-to-one correspondence between stochastic measures $\mu \in sca(\Sigma)$ and processes of integrable variation $V \in SV$. The main purpose of this section is to determine the class of processes $V \in SV$, which corresponds to the range of π'_x ($x = w, a, p$). This amounts to characterizing those processes of integrable variation, which commute with the projection π_x in the sense that they satisfy

$$E(\int Z_s \, dV_s) = E(\int (\pi Z)_x \, dV_s)$$

for every stochastic processes $Z = (Z_s) \in B^\infty(\Sigma)$.

Let us first characterize the measure, generated by an adapted increasing process $V \in SV$.

8.2 Lemma: Let μ be the real stochastic measure, generated by a process $V = (V_t) \in SV$. V is adapted if and only if $\mu([o, t] \times F) = \int E(X_F | F_t)$ $X_{[o,t]} \, d\mu$ holds for every $t \in R_+$ and $F \in F$.

Proof: The equality above reads $E(X_F V_t) = E(E(X_F | F_t) V_t)$ and the proof of the lemma follows from the fact that this equality holds for all $F \in F$ if and only if V_t is F_t measurable.

8.3 Proposition: A process $V = (V_t)$ of integrable variation is adapted if and only if the following equality holds for all bounded F-measurable functions f and every $t \in R_+$:

$$E(\int_{[o,t]} F_s \, dV_s) = E(\int f X_{[o,t]} \, dV_s) = E(F_t V_t) \, ;$$

where $F = (F_s)$ is a right continuous modification of the martingale

$$s \to E(f|F_s).$$

<u>Proof</u>: The sufficiency of the assertion is a consequence of Lemma 8.2.

Let $V = (V_t)$ be adapted. That $E(\int f X_{[0,t]} dV_s) = E(F_t V_t)$ holds is a consequence of Lemma 8.2 and the monotone class theorem.

In order to prove the remaining equality on the left hand side, put

$$Z^n = X_{[0]} F_0 + \Sigma_{k \geq 0} X_{]kt/n,\ (k+1)t/n]} F_{(k+1)t/n}$$

$$k = 0, 1, \ldots, n-1 \ ;\ n = 1, 2, \ldots$$

We may assume that all paths of the martingale $F = (F_s)$ are right continuous hence the sequence Z^n of stochastic processes converges pointwise to F and using the equality already proved, we get:

$$E(\int X_{[0,t]} Z^n dV_s) = \sum_{k=0}^{n-1} E[F_{(k+1)t/n}(V_{(k+1)t/n} - V_{kt/n})] + E(F_0 V_0)$$

$$= \sum_{k=1}^{n} E(F_{kt/n} V_{kt/n}) - \sum_{k=0}^{n-1} E(F_{kt/n} V_{kt/n}) + E(F_0 V_0)$$

$$= E(F_t V_t)$$

for every $n = 1, 2, \ldots$

The equality we want to prove is now a consequence of the Lebesgue dominated convergence theorem.

<u>Corollary</u>: A process of integrable variation $V = (V_t)$ is adapted if and only if for every bounded (resp. positive) right continuous martingale $M = (M_t)$ and for every stopping time T the following equality holds:

$$E(\int X_{[0,T]} M_t dV_t) = E(M_T V_T).$$

<u>Proof</u>: The sufficiency is obvious and in order to prove that the condition is necessary, apply Proposition 8.3 to the stopped processes $M^T = (M_{T \wedge t})$

71

Observing that every adapted stochastic process of bounded variation is well measurable, we get the following characterization of processes $V \in SV$, commuting with the well measurable projection π_w :

8.4 Theorem: A process $V = (V_t)$ of integrable variation commutes with the well measurable projection if and only if V is well measurable (adapted).

<u>Proof:</u> Notice that for a set $]s,t] \times F$, $(s \leq t$, $s,t \in R_+$, $F \in \mathcal{F})$, the well measurable projection of the process $X_{]s,t] \times F}$ is a right continuous modification of the process $u \to X_{]s,t]}(u)E(X_F|\mathcal{F}_u)$. In view of Proposition 8.3 the theorem is a consequence of the facts that $Z \to E(\int Z_u dV_u)$ defines a σ-additive measure on Σ and that sets of the form $]s,t] \times F$, $(s,t \in R_+$, $F \in \mathcal{F})$, are stable under finite intersections and generate the σ-algebra Σ.

As a consequence of Theorem 8.4 and Theorem 7.9 we get the following Corollary:

<u>Corollary:</u> Let $X = (X_t)$ and $Y = (Y_t)$ be two bounded measurable processes. The following assertions are equivalent:

(a) $E(X_T \chi_{\{T<\infty\}}) = E(Y_T \chi_{\{T<\infty\}})$ holds for every stopping time T.

(b) $E(\int X_{[0,T]} X_t dV_t) = E(\int X_{[0,T]} Y_t dV_t)$ holds for every stopping time T and every adapted process of integrable variation.

It is now easy to characterize processes of integrable variation commuting with the accessible projection:

8.5 Theorem: A process $V = (V_t)$ of integrable variation commutes with the accessible projection if and only if V is accessible.

Proof: If V commutes with the accessible projection then V commutes with the well measurable projection and is therefore adapted. Let T be a totally inaccessible stopping time. The indicator function $\chi_{[T]}$ of the graph of T has an accessible projection equal to zero, consequently $E(\int \chi_{[T]} dV_t) = 0$ holds for every totally inaccessible stopping time T. According to Proposition 7.14 V is accessible.

Suppose on the other hand V is accessible. In order to prove that V commutes with the accessible projection, it suffices to prove that V commutes with the accessible projection of every well measurable bounded process. For the same reasons as in the proof of the previous theorem it suffices to prove that

$$E(\int \chi_{[S,T[} dV_t) = E(\int \pi_a \chi_{[S,T[} dV_t)$$

holds for every process of the form $\chi_{S,T}$, where S and T are stopping times. Now $\chi_{[S,T[}$ and $\pi_a \chi_{[S,T[}$ differ only on the graph of a totally inaccessible stopping time, and the equality aobve holds because V doesn't charge totally inaccessible stopping times (Proposition 7.14).

Finally we have for predictable processes:

8.6 Theorem: A process $V = (V_t)$ of integrable variation commutes with the predictable projection if and only if V is predictable.

Proof: Notice that the predictable projection of a constant process χ_F ($F \in F$) is a left continuous modification of the martingale $t \to (E(\chi_F | F_{t-}))$. If V commutes with the predictable projection, then V is accessible and satisfies $E(\chi_F V_T) = E(E(\chi_F | F_{T-}) V_T)$ for every predictable stopping time T. This equality, being a consequence of the fact that the processes $\chi_{]0,T]} \chi_F$ and $\chi_{]0,T]} E(\chi_F | F_{T-})$ have the same predictable projection, proves according

73

to Theorem 7.8 that V is predictable.

The proof of the sufficiency is a consequence of the following:

8.7 Lemma: Let $V = (V_t)$ be a predictable increasing process. For every bounded F-measurable function f we have $E(\int_{[0,t]} F_s dV_s) = E(fV_t)$, where $F = (F_s)$ is a left continuous version of the martingale $(E(f|F_{t-}))$.

The proof of this Lemma is similar to the proof of Proposition 7.2. Notice that Theorem 7.8 yields $E(fV_s) = E[E(f|F_{s-})V_s]$ for every $s \in R_+$. Then approximate $F = (F_s)$ by the sequence (Y^n) of processes

$$Y^n = \sum_{k=0}^{n-1} \chi_{[kt/n,(k+1)t/n[} F_{kt/n} \quad ; \quad n = 1,2,\ldots$$

The proof of the sufficiency part of Theorem 8.6 is now analogous to the proof in the well measurable case.

Corollary: Let $V = (V_t)$ be an adapted increasing process. V is predictable if and only if

$$E(\int_{[0,t]} M_s dV_s) = E(\int_{[0,t]} M_{s-} dV_s)$$

holds for every $t \in R_+$ and every right continuous bounded martingale $M = (M_t)$.

Proof: The fact that the predictable projection of the martingale (M_t) is the process (M_{t-}) proves the necessity. The sufficiency is proved as the first part of Theorem 8.6.

Remark: The condition in the Corollary above defines natural increasing processes (see [19]).

8.8: Let us now summarize and apply the results obtained so far in this section. Denote by ϕ the isometric isomorphism which maps a process of integrable variation $V = (V_t) \in SV$ to the stochastic measure $\mu_V \in sca(\Sigma)$, generated by V. μ_V is determined by the equality $\int Z d\mu_V = E(\int Z_t dV_t)$ for every measurable bounded process $Z = (Z_t) \in B^\infty(\Sigma)$.

Write wSV (resp. aSV, pSV) for the normed space of well measurable (resp. accessible, predictable) processes of integrable variation.

Furthermore we write i for the imbedding map of xSV into SV and q for the imbedding map of $B^\infty(\Sigma_x)$ into $B^\infty(\Sigma)$. The adjoint q' of q maps stochastic measures $\mu \in sca(\Sigma)$ to their restriction to the σ-algebra Σ_x $x = (w, a, p)$.

The following diagram illustrates the situation:

$$\begin{array}{ccccc}
B^\infty(\Sigma) & \xrightarrow{\pi_x} & B^\infty(\Sigma_x) & \xrightarrow{q} & B^\infty(\Sigma) \\
& & & & \\
sca(\Sigma) & \xleftarrow{\pi_x'} & sca(\Sigma_x) & \xleftarrow{q'} & sca(\Sigma) \\
\downarrow \phi^{-1} & & \downarrow \phi^{-1} \circ \pi_x' & & \downarrow \phi^{-1} \\
SV & \xleftarrow{i} & xSV & \xleftarrow{\phi^{-1} \circ \pi_x' \circ q' \circ \phi} & SV
\end{array}$$

A stochastic process $V \in SV$ is an element of xSV if and only if

$$E(\int (\pi_x Z)_t dV_t) = E(\int Z_t dV_t)$$

is satisfied for every bounded measurable process $Z = (Z_t) \in B^\infty(\Sigma)$. The map $\phi^{-1} \circ \pi_x'$ establishes an isometric (Banach lattice) isomorphism between $sca(\Sigma_x)$ and xSV.

For an element $V \in SV$ we call the process $\phi^{-1} \circ \pi_x' \circ q' \circ \phi(V)$ the dual Σ_x-measurable projection of V, and we shall write in the sequel simply $\pi_x' V$ instead of $\phi^{-1} \circ \pi_x' \circ q' \circ \phi(V)$. The dual well measurable (resp. accessible, predictable) projection of a process of integrable variation $V \in SV$ is the unique process $W \in wSV$ (resp. aSV, pSV) which satisfies $\mu_W(A) = \mu_V(A)$ for elements $A \in \Sigma_w$ (resp. $A \in \Sigma_a$, $A \in \Sigma_p$)

The following proposition will yield two remarkable properties of the Σ_x-measurable projection:

8.9 Proposition: Let $V = (V_t)$ be an element of SV ($x = w, a, p$) and $Z = (Z_t) \in B^\infty(\Sigma)$. For every bounded Σ_x-measurable process $Y = (Y_t) \in B^\infty(\Sigma_x)$ the equality $\int Y \pi_x Z d\mu_V = \int \pi_x(YZ) d\mu_V = \int YZ d\mu_{\pi'_x V}$ holds.

Proof: The equality on the left hand side is a consequence of Theorem 7.15 (c) and the equality on the right hand side follows immediately from the definition of the dual Σ_x-measurable projection.

Corollary 1: The equality $\pi'_x V(Z) = [\pi'_x V](Z)$ holds for every element $V \in SV$ and $Z \in B^\infty(\Sigma_x)$; $x = w, a, p$.

Corollary 2: The equality $\pi'_x V(Z) = V(\pi_x Z)$ holds for every element $V \in xSV$ and $Z \in B^\infty(\Sigma)$; $x = w, a, p$.

Remark: Recall that $V(Z) = (V(Z)_t)$ is the stochastic integral of Z with respect to V (see 2.2).

Proof: We prove Corollary 2. In order to prove the equality it suffices to prove the corresponding equality for the measures $\mu_{V(Z)}$ and $\mu_{V(\pi_x Z)}$, and this equality follows immediately from the previous proposition.

We conclude this section by studying right continuous martingales of integrable variation, i.e. stochastic processes $V \in wSV$ which are martingales.

8.10 Proposition: Let $V = (V_t)$ be an adapted stochastic process of integrable variation, which is zero at time $t = o$. The following assertions are equivalent:

(a) V is a martingale.

(b) The dual predictable projection of V is evanescent.

(c) μ_V restricted to Σ_p is zero.

<u>Proof</u>: The equivalence of (b) and (c) is clear after the remark following the definition of the dual predictable projection (8.8).

In order to prove the equivalence between (a) and (c), recall that the predictable σ-algebra Σ_p is generated by stochastic intervals of the form $[0_F]$ $(F \in F_0)$ and $]s_F, t_F] =]s,t] \times F$ $(s \leq t, F \in F_s)$ according to 7.4. Now $\mu_V([0_F]) = o$ holds because V was assumed to be o at time $t = o$ and $E(V_t - V_s | F_s) = o$ for every $t \geq s$ is equivalent to $E[\chi_F(V_t - V_s)] = \mu_V(]s_F, t_F]) = o$ for every $t \geq s$, $F \in F_s$.

<u>Corollary 1</u>: If a predictable process V of integrable variation is a martingale, the V is constant.

<u>Corollary 2</u>: Let $M = (M_t)$ and $V = (V_t)$ be two right continuous martingales, satisfying

(a) V is a process of integrable variation.

(b) M is uniformly integrable and the constant process
$$Z = \sup_{s \in R_+} |M_s|$$
is integrable with respect to V.

Then $E(\int M_s dV_s) = E(M_\infty V_\infty) = E(\Sigma_s \Delta M_s \Delta V_s)$ holds, where $\Delta M_s = M_s - M_{s-}$ and $\Delta V_s = V_s - V_{s-}$ and $M_{0-} = V_{0-} = o$.

<u>Proof</u>: (b) implies that $M = (M_s)$ and $M_- = (M_{s-})$ are integrable with respect to V. We have $E(M_\infty V_\infty) = E(\int M_s dV_s)$ by 8.3 Corollary. Now μ_V restricted to Σ_p is o and the process M_- is predictable, hence
$$E(\int M_s dV_s) = E(\int (M_s - M_{s-}) dV_s) = E(\Sigma_s \Delta M_s \Delta V_s)$$
holds.

Corollary 3: Under the hypotheses of Corollary 2

$$M_t V_t - \Sigma_{s \leq t} \Delta M_s \Delta V_s$$

is a uniformly integrable martingale, zero at time $t = 0$.

Proof: $(\sup_{s \in R_+} |M_s|)|V|_\infty$ is integrable according to b) and

$$|M_t V_t| \leq (\sup_{s \in R_+} |M_s|)|V|_\infty \quad \text{and} \quad |\Sigma_{s \leq t} \Delta M_s \Delta V_s| \leq 2(\sup_{t \in R_+} |M_s|)|V|_\infty$$

holds for every $t \in R_+$, hence the uniform integrability. In order to prove that the difference above is a martingale, apply Corollary 2 to the stopped processes M^T and V^T for every stopping time T.

Corollary 4: Let $V = (V_t)$ be an adapted process of integrable variation and denote by $W = (W_t)$ the dual predictable projection of V. $V - W$ is a martingale. If V is of the form $V = f\chi_{[T, \infty[}$, where f is an F_T-measurable integrable function and T is a stopping time, then W is a continuous stochastic process in the case that T is totally inaccessible and W is of the form $W = E(f|F_{T-})\chi_{[T, \infty[}$ in the case that T is a predictable stopping time.

Proof: The first part of Corollary 4 is obvious from Proposition 8.10. In order to see that W is continuous for the case of a totally inaccessible stopping time T, notice that the predictable W cannot charge a totally inaccessible stopping time according to 7.13 Corollary 1, and since the measures, generated by V and W, coincide on Σ_p, W cannot charge a predictable stopping time. Hence W must be continuous.

In the case of a predictable stopping time T, notice that $W = E(f|F_{T-})\chi_{[T, \infty[}$ is a predictable process according to Proposition 7.14. Furthermore the real measures, μ_V and μ_W, generated by V respective W

are carried by the graph $[T]$ of the stopping time T. Now the predictable subsets of T are of the form $(R_+ \times F) \cap [T]$ $(F \in F_{T-})$ according to 7.6 Corollary 2 and Proposition 6.8, hence the equality

$$\int \chi_F \chi_{[T]} d\mu_V = E(\chi_F f) = E(\chi_F E(f|F_{T-})) = \int \chi_F \chi_{[T]} d\mu_W$$

proves the assertion $\mu_V = \mu_W$ on Σ_p.

§9 QUASIMARTINGALES

In this section a subspace of the vector space of real stochastic processes will be studied, the elements of which are called quasimartingales. It is easy to see that every right continuous $L^1(P)$-bounded supermartingale is a quasimartingale (for the definition see below). On the other hand, the family of right continuous $L^1(P)$-bounded supermartingales form a cone in the vectorspace of real stochastic processes, and the space of quasimartingales will turn out to be the subspace, generated by this cone.

We shall associate to every quasimartingale a bounded additive measure on a Boolean ring of subsets of $R_+ \times \Omega$. Decompositions of these measures will give rise to corresponding decompositions of quasimartingales.

As usual, we fix a stochastic base

$$(\Omega, F, P, (F_t), R_+).$$

Note that $\|f\|_1$ means the norm $f \to E(|f|)$ on $L^1(P)$.

9.1 Definition: An $L^1(P)$-bounded adapted right continuous stochastic process $X = (X_t)$ is called a *quasimartingale*, if there exists a constant $K \geq 0$, such that

$$\sum_{i=1}^{n} \|E(X_{t_{i+1}} - X_{t_i} | F_{t_i})\|_1 \leq K$$

holds for every finite increasing sequence $t_1, t_2, \ldots, t_{n+1}$ of elements of R_+.

Examples: (1) Let (M_t) and $N = (N_t)$ be two square integrable right continuous martingales satisfying $\sup_t \|M_t\|_2 \leq K_1$ and $\sup_t \|N_t\|_2 \leq K_2$ respectively. Then the process $X = (X_t)$, defined by $X_t = M_t N_t$, is a quasimartingale. Indeed, for an increasing sequence $t_1 < t_2 \ldots < t_{n+1}$ of elements of R_+ we have $\sum_i \|E(M_{t_{i+1}} N_{t_{i+1}} | F_{t_i}) - M_{t_i} N_{t_i}\|_1$

$$= \Sigma_i \|E[(M_{t_{i+1}} - M_{t_i})(N_{t_{i+1}} - N_{t_i})|F_{t_i}]\|_1$$

$$\leq \Sigma_i \|(M_{t_{i+1}} - M_{t_i})(N_{t_{i+1}} - N_{t_i})\|_1$$

$$\leq (\Sigma_i \|M_{t_{i+1}} - M_{t_i}\|_2^2)^{1/2} (\Sigma_i \|N_{t_{i+1}} - N_{t_i}\|_2^2)^{1/2}$$

$$= \|M_{t_{n+1}} - M_{t_1}\|_2 \|N_{t_{n+1}} - N_{t_1}\|_2$$

$$\leq 4 K_1 K_2$$

according to Holder's inequality and Schwarz inequality.

(2) Let $X = (X_t)$ be a right continuous stochastic process defined on a base (Ω, F_o, P) and let us assume that X has independent increments, i.e. that X satisfies the following condition: for every finite sequence $0 = t_o < t_1 < \ldots < t_n$ of elements of R_+ $\{X_{t_o}, X_{t_1} - X_{t_o}, \ldots, X_{t_n} - X_{t_{n-1}}\}$ is an independent family of random variables. For an element $t \in R_+$ let E_t be the σ-algebra generated by the family of random variables $\{X_s : s \leq t\}$ and denote by F the completion of the σ-algebra $\bigvee_{t \in R_+} E_t$. If we write F_t for the σ-algebra E_{t+} augmented by the P-null sets of F, then $X = (X_t)$ is an adapted process with respect to the stochastic base $(\Omega, F, P, (F_t), R_+)$. For a sequence $t_o, t_1, \ldots, t_n$ of elements of R_+ we denote by $E(X_t | X_{t_o}, X_{t_1}, \ldots, X_{t_n})$ the conditional expectation of X_t with respect to the σ-algebra generated by the family of random variables $\{X_{t_o}, X_{t_1}, \ldots, X_{t_n}\}$.

For an arbitrary sequence $t_o, \ldots, t_n$ of elements of R_+ with $0 = t_o < \ldots < t_n < t$ we have the equality

$$E(X_t | X_{t_o}, \ldots, X_{t_n}) = E(X_t | X_{t_o}, X_{t_1} - X_{t_o}, \ldots, X_{t_n} - X_{t_{n-1}})$$

$$= E(X_t - X_{t_n} | X_{t_o}, X_{t_1} - X_{t_o}, \ldots, X_{t_n} - X_{t_{n-1}}) + E(X_{t_n} | X_{t_o}, \ldots, X_{t_n})$$

$$= E(X_t - X_{t_n}) + X_{t_n}.$$

From this equality we deduce that the equality

$$E(X_t | F_s) = E(X_t - X_s) + X_s$$

holds for every pair s,t of elements of R_+ with $s < t$, provided all the random variables X_t are integrable. Consequently $X = (X_t)$ is a quasimartingale if and only if the function $t \to E(X_t)$ is of bounded variation on R_+. Suppose now that $X = (X_t)$ is a quasimartingale, and define the process $Y = (Y_t)$ by $Y_t = E(X_t)$ $(t \in R_+)$. Then $X - Y$ is a martingale and Y is a predictable process of integrable variation. One of the objectives of this section is to obtain a similar decomposition for arbitrary quasimartingales.

For the rest of this section, we define for every stochastic process $X = (X_t)$ the random variable X_∞ by $X_\infty = 0$. But we refer to X as a (super-, sub-) martingale resp. as an increasing process if $(X_t)_{t \in R_+}$ has the properties in question.

9.2 Theorem: Let $X = (X_t)$ be a right continuous adapted process. The following assertions are equivalent:

(a) X is a quasimartingale.

(b) There is a constant $K \geq 0$, such that for every finite increasing sequence $S_1, \ldots, S_n$ of stopping times $\Sigma_i |E(X_{S_{i+1}} - X_{S_i})| \leq K$ holds.

(c) There is a constant $K \geq 0$, such that for every finite increasing sequence $S_1, \ldots, S_n$ of stopping times

$\Sigma_i \|E(X_{S_{i+1}} - X_{S_i}|F_{S_i})\|_1 \leq K$ holds.

Proof: Let us first prove a slightly different version of the theorem above, replacing 'stopping times' in the assertions (b) and (c) by 'simple stopping times'.

(a) ⇒ (b) : Denote by $t_1 < t_2 < \ldots < t_m$ the values taken by the family $S_1, \ldots, S_n$ of simple stopping times and write $\Delta_0 = t_1$, $\Delta_k = t_{k+1} - t_k$ for $k = 1, \ldots, m - 1$. (Note that t_m can be equal to ∞.)

Every stopping time S_i can be written in the form

$$S_i = \Sigma_{k=0}^{m-1} \chi_{A_k^i} \Delta_k ,$$

where $A_k^i = \{S_i \geq t_k\}$ is an element of F_{t_k} and $A_k^1 \subseteq A_k^2 \subseteq \ldots \subseteq A_k^n$ holds for every fixed k.

We have, putting $X_{t_0} = 0$

$$\Sigma_{i=1}^{n-1} |E(X_{S_{i+1}} - X_{S_i})|$$

$$= \Sigma_{i=1}^{n-1} |E[\Sigma_{k=0}^{m-1} (\chi_{A_k^{i+1}} - \chi_{A_k^i})(X_{t_{k+1}} - X_{t_k})]|$$

$$\leq E[\Sigma_{k=0}^{m-1} \Sigma_{i=1}^{n-1} (\chi_{A_k^{i+1}} - \chi_{A_k^i}) E(X_{t_{k+1}} - X_{t_k}|F_{t_k})|]$$

$$\leq \Sigma_{k=1}^{m-1} \|E(X_{t_{k+1}} - X_{t_k}|F_{t_k})\|_1 + \sup_{t \in R_+} \|X_t\|_1$$

$$\leq K$$

for a constant K, because X is a quasimartingale.

(b) ⇒ (c) : Let $R_1, \ldots, R_n$ be an increasing family of simple stopping times. Put $A_k = \{E(X_{R_{k+1}} - X_{R_k}|F_{R_k}) \geq 0\}$ and define the stopping times U_k resp. V_k as the restrictions of R_k resp. R_{k+1} to A_k.

We have (9.2.1):

$$\Sigma_k E([E(X_{R_{k+1}} - X_{R_k} | F_{R_k})]^+) = \Sigma_k E(X_{V_k} - X_{U_k}),$$

Denote by S_1 the debut of the predictable set $A_1 = \bigcup_k U_k, V_k$, by T_1 the debut of A_1^c, define S_2 as debut of $A_2 = A_1 \cap T_1, \infty$, T_2 as debut of A_2^c, and so on.

There exists a positive integer m, such that T_m is equal to ∞, and for the sequence $S_1 \leq T_1 \leq S_2 \leq T_2 \leq \ldots \leq T_m$ of simple stopping times we have:

$$\sum_{i=1}^{m} (X_{T_i} - X_{S_i}) = \sum_{k=1}^{n} X_{V_k} - X_{U_k}.$$

Together with (9.2.1) this gives

$$\Sigma_{k=1}^{n} E([X_{R_{k+1}} - X_{R_k} | F_{R_k}]^+) \leq K$$

and using the same argument for the negative part, we get finally

$$\Sigma_k E[|E(X_{R_{k+1}} - X_{R_k} | F_{R_k})|] \leq 2K.$$

That c) implies a) is obvious.

Taking into account that every stopping time T is the limit of a decreasing sequence of simple stopping times, the proof of the theorem is a consequence of the following.

9.3 Lemma: Let $X = (X_t)$ be a quasimartingale and (T_n) a decreasing sequence of simple stopping times T_n with $T = \lim_n T_n$. The sequence (X_{T_n}) of random variables converges to X_T in the $L^1(P)$-norm.

Proof: Let an $\varepsilon > 0$ be given. According to the version of the previous theorem, which is already proved (a) $\Rightarrow$ c)), there exists a positive integer k, such that $\|E(X_{T_k} - X_{T_n} | F_{T_k})\|_1 < \varepsilon$ holds for all $n \geq k$.

Define the reverse martingale $M_n = E(X_{T_k} | F_{T_n})$, $(n \geq 1)$. According to Proposition 4.14 the family (M_n) is uniformly integrable, i.e. there exists a $\delta > 0$, such that $A \in F$ and $P(A) < \delta$ implies $E(\chi_A |M_n|) < \varepsilon$, for all n. Now the sequence (X_{T_n}) converges P-almost surely to X_T, consequently there exists a k', such that $P\{|X_{T_n} - X_{T_m}| > \varepsilon\} < \delta$ holds for every $m,n \geq k'$. Put $q = \max\{k, k'\}$ and let for $m,n \geq q$ A be the set $A = \{|X_{T_n} - X_{T_m}| > \varepsilon\}$.

We have

$$E(|X_{T_n} - X_{T_m}|) \leq$$

$$\leq \varepsilon + E(\chi_A |X_{T_n} - X_{T_m}|)$$

$$\leq \varepsilon + E(\chi_A |M_n|) + E(\chi_A |M_m|) +$$

$$+ E(\chi_A |X_{T_n} - E(X_{T_k} | F_{T_n})|) +$$

$$+ E(\chi_A |X_{T_m} - E(X_{T_k} | F_{T_m})|)$$

$$\leq 5\varepsilon$$

for all $m,n \geq q$.

That proves, that (X_{T_n}) is a Cauchy sequence in $L^1(P)$. Since (X_{T_n}) takes place in the $L^1(P)$-norm.

Having now established the general form of Theorem 9.2, the proof above applies without change to the following reinforced version of Lemma 9.3:

9.4 Proposition:

Let $X = (X_t)$ be a quasimartingale and (T_n) a decreasing sequence of stopping times with $T = \lim_n T_n$. Then X_{T_n} is integrable for every n and the sequence (X_{T_n}) converges to X_T P-almost surely and in the

$L^1(P)$-norm.

The following proposition states a conditional form of 9.2 c):

9.5 Theorem:

A right continuous adapted process $X = (X_t)$ is a quasimartingale if and only if for every stopping time T there exists a (F_T-measurable) positive integrable function $f \in L^1(P)$, such that $\Sigma |E(X_{S_{i+1}} - X_{S_i} | F_T)| \leq f$ holds for every finite increasing sequence $S_1, \ldots, S_n$ of stopping times, satisfying $S_1 \geq T$.

Proof: By writing the condition above for $T = 0$ and taking expectations, it is easily seen that the condition implies 9.2 b).

In order to prove the necessity, we observe that according to 9.2 c) $\Sigma \| E(X_{S_{i+1}} - X_{S_i} | F_T) \|_1 \leq K$ holds for every sequence of stopping times, satisfying the hypothesis. From the monotone convergence theorem we therefore deduce that for every increasing sequence (S_j) of stopping times $S_j \geq T$ the sum $\Sigma_j |E(X_{S_{j+1}} - X_{S_j} | F_T)|$ is convergent to an element $f \in L^1(P)$, and the necessity of the condition follows from the fact that a subset of $L^1(P)$ has an upper bound $f \in L^1(P)$, if and only if every countable subset has.

9.6: We now have enough information about quasimartingales to study the associated measure, which will be defined on a ring R of subsets of $R_+ \times \Omega$. A *stochastic interval* $]S,T]$ is called *bounded*, if $]S,T]$ is contained in a stochastic interval of the form $]0,t]$ for an element $t \in R_+$. (For notation see §6). The Boolean ring R is defined as the ring generated by sets of the form $[0_F]$ ($F \in \bigcup_{t \in R_+} F_t$) or $]S,T]$ where $]S,T]$ is a *bounded* stochastic interval.

Note that the intersection of R with the predictable σ-algebra Σ_p is a ring, generating Σ_p, and that a measure μ of bounded variation, which is

σ-additive on $R \cap \Sigma_p$ has a unique extension to a σ-additive measure on Σ_p.

It is clear that every element H of R can be written uniquely as $H = [0_F] \cup]S_1, T_1] \cup \ldots \cup]S_n, T_n]$, where F is an element of the Boolean algebra $\bigcup_{t \in R_+} F_t$ and $]S_j, T_j]$ $(j=1, \ldots, n)$ are stopping times, satisfying $S_j \leq T_j \leq S_{j+1}$ and $S_j \neq T_j \neq S_{j+1}$ for $j = 1, 2, \ldots, n$.

Next notice that, given a quasimartingale $X = (X_t)$, for every $t \in R_+$ and $F \in F_t$ the limit $\lim_{h \to \infty} E(\chi_F X_{t+h})$ exists according to 9.2 b).

For an element $H \in R$, which is the disjoint union
$$H = [0_F] \cup]S_1, T_1] \cup \ldots \cup]S_n, T_n] \quad (F \in F_t)$$
we define the *measure* μ_X *associated with the quasimartingale*
$$X = (X_t) \text{ as } \mu_X(H) = \sum_{j=1}^{n} E(X_{T_j} - X_{S_j}) - \lim_{h \to \infty} E(\chi_F X_{t+h}).$$

The properties a) and b) are easy consequences of the definitions and the martingale convergence theorem:

a) X is a martingale if and only if μ_X is carried by the set $[0] \in R$, X is a uniformly integrable martingale if and only if μ_X is carried by $[0]$ and is moreover σ-additive.

b) X is a positive supermartingale if and only if μ_X is negative.

The assertion c) is a consequence of Theorem 9.5 and the definition of μ_X:

c) For every $t \in R_+$ there exists a positive element $f \in L^1(P)$, such that $|\mu_X|(]t_F, \infty]) \leq E(\chi_F f)$ holds for every $F \in F_T$.

(Notice that the total variation $|\mu_X|(A)$ is defined for sets A not necessarily in R, $|\mu_X|$ is however in general an additive set function only on R).

c) together with the definition of μ_X implies:

d) the set function $F \to \mu_X([O_F])$ on F_t is a P-absolutely continuous (σ-additive) measure on F_t for every $t \in R_+$.

The assertion e) finally is a consequence of Proposition 9.4 and Theorem 9.2:

e) μ_X is of bounded variation and satisfies $\lim_{n \to \infty} \mu_X(]T,T_n]) = 0$ for every decreasing sequence $(]T,T_n])$ of stochastic intervals $]T,T_n] \in R$ with $\cap_n]T,T_n] = \emptyset$.

In the sequel we shall say that a measure μ on R satisfies' *condition (S)'*, if μ satisfies the assertions c), d) and e).

Note that if the assertions c) and d) hold for a measure μ on R, every evanescent subset A of $R_+ \times \Omega$, which is contained in $[0,t]$ for a $t \in R_+$, is a μ-null set.

The significance of condition (S) becomes clear from the following:

9.7 Proposition: Let μ be a negative measure on R, satisfying (S). There exists an (up to indistinguishability) unique right continuous supermartingale $X = (X_t)$, such that the associated measure μ_X coincides with μ.

Proof: For $t \in R$ and $F \in F_t$ define

$$\mu^t(F) = - \lim_{h \to \infty} \mu(]t_F, (t+h)_F]) - \mu([O_F]).$$

According to c) and d) of (S) $F \to \mu^t(F)$ is a P-absolutely continuous finite positive measure on F_t. Denote by X_t the Radon-Nikodym derivative $X_t = d\mu^t/dP$. (X_t) is a positive supermartingale and has a right continuous modification according to Proposition 5.1 and 5.3 Corollary 1, because μ

satisfies e), which implies that $t \to E(X_t)$ is right continuous.

We shall use this one-to-one correspondence between measures satisfying (S) and stochastic processes to decompose quasimartingales via a decomposition of the associated measures. The family $ba(R)$ of all measures of bounded variation on R forms an order complete vector lattice under the natural ordering. The decompositions $\mu = \mu_1 + \mu_2$, we shall use in the sequel, will always have the property that if μ is an element of a lattice ideal I, then μ_1 and μ_2 are elements of I.

(A lattice ideal of a vector lattice L is a sub-vector lattice I of L, which satisfies: $\mu \in L$ and $0 \leq \mu \leq \nu$ for an element ν of $I \Rightarrow \mu \in I$. Examples for lattice ideals of $ba(R)$ are: $ca(R)$, the family of σ-additive measures $\mu \in ba(R)$; the family of purely additive measures $\mu \in ba(R)$, i.e. the subspace of $ba(R)$, (lattice-) orthogonal to $ca(R)$; any family of measures $\mu \in ba(R)$, which are carried by a (fixed) subset $H \in R$).

In order to make sure that we remain after decomposition in the class $sba(R)$ of elements $\mu \in ba(R)$ satisfying (S), we shall first prove the following:

9.8 Proposition: The family $sba(R)$ of measures $\mu \in ba(R)$, satisfying (S), is a lattice ideal in $ba(R)$.

<u>Proof</u>: It is clear that the family I_1 of all $\mu \in ba(R)$, which satisfy c) and d) of condition (S) form a lattice ideal.

Let us prove that the family I_2 of all $\mu \in ba(R)$, satisfying e) of condition (S) is a sub-vector lattice. We have to show that $\mu \in I_2$ implies that $|\mu|$ satisfies e) of (S) or equivalently satisfies $\lim_{n \to \infty} |\mu|(]T_n, T_0]) = |\mu|(]T, T_0])$ for every increasing sequence

$(]T_n,T_o])$ of stochastic intervals $]T_n,T_o] \in R$ with $\bigcup_n]T_n,T_o] =]T,T_o]$. We certainly have $\lim_{n\to\infty} |\mu|(]T_n,T_o]) \leq |\mu|(]T,T_o])$. Now notice that for every element $H \in R$ $\lim_{n\to\infty} \mu(]T_n,T_o] \cap H) = \mu(]T,T_o] \cap H)$ is valid by hypothesis, hence for a partition of $R_+ \times \Omega$ into disjoint elements $H_1,\ldots,H_m$ of R we get:

$$\Sigma_k |\mu(]T,T_o] \cap H_k)| = \Sigma_k |\lim_n \mu(]T_n,T_o] \cap H_k)|$$
$$= \lim_n (\Sigma_k |\mu(]T_n,T_o] \cap H_k|)$$
$$\leq \lim_n |\mu|(]T_n,T_o]).$$

This proves that $|\mu|(]T,T_o]) \leq \lim_n |\mu|(]T_n,T_o])$ holds too, hence I_2 is a sublattice of $ba(R)$. That I_2 is in fact a lattice ideal is now an obvious consequence of e).

Finally $sba(R)$ is the intersection of I_1 and I_2 and therefore a lattice ideal.

Decomposing μ_X into its positive part and negative part, the following theorem is an immediate consequence of the propositions 9.8 and 9.7.

9.9 Theorem: Every quasimartingale can be written as a difference of two positive right continuous supermartingales.

The following corollary extends Proposition 9.7:

<u>Corollary 1</u>: Let $\mu \in sba(R)$ be a measure on R satisfying (S). There exists an (up to indistinguishability) unique quasimartingale $X = (X_t)$, such that the associated measure μ_X coincides with μ.

As an immediate consequence of the supermartingale - convergence theorem we have:

<u>Corollary 2</u>: Let $X = (X_t)$ be a quasimartingale. The random varibles X_t

converge P-almost surely to an integrable random variable X_∞^- as t tends to ∞.

A quasimartingale $X = (X_t)$ is called a *quasipotential*, if $\lim_{t \to \infty} X_t = 0$ holds in the $L^1(P)$-norm.

If X is moreover a positive supermartingale, then X is called simply a *potential*.

It is obvious that a positive supermartingale is a potential if and only if $[0]$ is a μ_X-null set. Since μ_X-null sets are nullsets of the positive part and negative part of μ_X, we have:

9.10 Proposition: X is a quasipotential if and only if $[0]$ is a μ_X-null set.

Decomposing μ_X into the sum $\mu_X = \mu_M + \mu_P$, where μ_M is the restriction of μ_X to the set $[0]$, we get the *Riesz decomposition* of quasimartingales:

9.11 Theorem: Every quasimartingale $X = (X_t)$ can be written as sum $X = M + P$ of a right continuous martingale $M = (M_t)$ and a quasipotential $P = (P_t)$. This decomposition is unique.

Proof: M (resp. P) are the quasimartingales corresponding to μ_M resp. to μ_P in the decomposition $\mu_X = \mu_M + \mu_P$ above. That M is a martingale follows from 9.6 a) and that P is a quasipotential from Proposition 9.10.

The uniqueness is an easy consequence of the martingale convergence theorem.

If μ_X is σ-additive, a different interpretation of μ_P gives us the *Doob-Meyer decomposition* of quasimartingales:

9.12 Theorem: Let $X = (X_t)$ be a quasimartingale such that μ_X is σ-additive on R. X has a unique decomposition $X = M + V$ into a uniformly integrable right continuous martingale $M = (M_t)$ and a predictable process of integrable variation $V = (V_t) \in pSV_o$, zero at time $t = 0$.

Proof: The uniqueness of the decomposition is clear in view of 8.10 Corollary 1. Let us prove the existence. Decompose μ_X into $\mu_X = \mu_M + \mu_P$ as in the Riesz decomposition and observe that both μ_M and μ_P are σ-additive. In particular μ_P is a σ-additive bounded measure on $R \cap \Sigma_p$, where Σ_p is the predictable σ-algebra, hence μ_p has a unique extension to a σ-additive measure $\bar{\mu}_p$ on Σ_p and $\bar{\mu}_p$ is a real stochastic measure on Σ_p, because c) of condition (S) holds for μ_p.

According to 8.8 there exists a unique process $V = (V_t)$ of integrable variation, (zero at time $t = 0$, because $[0]$ is a μ_p-null set), generating $\bar{\mu}_p$, i.e. satisfying $E(V_T - V_S) = \mu_p(]S,T])$ for every pair S,T of stopping times, such that $]S,T]$ is a stochastic interval contained in R.

It is obvious that the measure μ_V, associated to V (in the sense of this section) decomposes into $\mu_V = \mu_{M'} + \mu_{P'}$, where $\mu_{M'}$ is carried by $[0]$ and defined as $\mu_{M'}([0_F]) = -E(\chi_F V_\infty^-)$. Hence μ_X has a decomposition $\mu_X = (\mu_M - \mu_{M'}) + \mu_V$ into σ-additive measures on R, and if we denote by $M = (M_t)$ the martingale associated to $(\mu_M - \mu_{M'})$, $X = M + V$ is the desired decomposition of X.

A necessary and sufficient condition for μ_X to be σ-additive is given in the following.

9.13 Proposition: Let $X = (X_t)$ be a quasimartingale. The corresponding measure μ_X is σ-additive if and only if X is of class (D), i.e. satisfies the following condition: The set $\{X_T : T$ is a finite stopping time$\}$ is uniformly integrable.

Remark: The quasimartingale $X = (X_t)$ with $X_t = M_t N_t$ of example 1) (following Definition 9.1) is a quasimartingale of class (D): If we put $M^* = \sup_t |M_t|$ and $N^* = \sup_t |N_t|$, then M^* and N^* are elements of $L^2(P)$ according to Theorem 4.11. Therefore the product $M^* N^*$ is integrable and, since for every finite stopping time T the inequality $|X_T| \leq M^* N^*$ holds, the process $X = (X_t)$ is of class (D).

Proof: If μ_X is σ-additive, X is the sum of a uniformly integrable martingale and a process of integrable variation according to Theorem 9.11. Both processes are of class (D) and consequently X is of class (D).

Let now X be of class (D). We first observe that the convergence of X_t to X_∞ (9.8 Corollary 2) takes place in the $L^1(P)$-norm in this case, hence the measure μ_X restricted to the set $[0] \in R$ is σ-additive. We therefore may assume that $[0]$ is a μ_X-null set. For an element $A = [0_F] \cup]S_1, T_1] \cup \ldots \cup]S_k, T_k]$ we write $\bar{A}$ for the set $\bar{A} = [0_F] \cup [S_1, T_1] \cup \ldots \cup [S_k, T_k]$. In view of e) of condition (S) (see 9.6) and Proposition 9.8, for every $\delta > 0$ and $A \in R$, we can find an element $H \in R$ with $\bar{H} \subseteq A$, such that $|\mu_X|(A) \leq |\mu_X|(H) + \delta$ holds for the modulus $|\mu_X|$ of μ_X.

Let now (A_n) be a decreasing sequence of elements of R with $\bigcap_n A_n = \emptyset$. We may assume that A_n is contained in a stochastic interval $[0, t]$ ($t \in R_+$). For every $\varepsilon > 0$ there exists a partition $B_1, \ldots, B_k$ of $[0, t]$ into elements B_j of the form $]U_j, V_j]$ such that $|\mu_X|([0, t[) \leq \sum_j |\mu_X(B_j)| + \varepsilon$ holds and in particular we have for every set $B \subseteq B_j$:

$$|\mu_X|(B) = |\mu_X|(B_j) - |\mu_X|(B_j \setminus B)$$

$$\leq |\mu_X(B_j)| + \varepsilon_j - |\mu_X(B_j \setminus B)|$$

$$\le |\mu_X(B)| + \varepsilon_j \quad ; \quad (9.13.1)$$

where $0 \le |\mu_X|(B_j) - |\mu_X(B_j)| = \varepsilon_j$ and $\Sigma_j \varepsilon_j = \varepsilon$ is valid.

In order to show that $\lim_n |\mu_X|(A_n) = 0$ holds, it suffices to show that for every $\varepsilon > 0$ $\lim_n |\mu_X|(A_n \cap B_j) \le \varepsilon + \varepsilon_j$ is valid for every j, where $B_1, \ldots, B_k$ is the decomposition above.

There exists an element $H_n \in R$, such that $\bar{H}_n \subseteq A_n \cap B_j$ and $|\mu_X|(A_n \cap B_j) \le |\mu_X|(H_n) + 2^{-n}\varepsilon$ holds for every n. Put $G_n = H_1 \cap H_2 \cap \ldots \cap H_n$. Then for every n $|\mu_X|(A_n \cap B_j) \le |\mu_X|(G_n) + \varepsilon$ holds. Now notice that the set $\bar{G}_n$ contains the graph $[D_n]$ of its debut D_n and since $(\bar{G}_n)$ is a decreasing sequence of well measurable subsets of $R_+ \times \Omega$ with $\cap_n \bar{G}_n = \emptyset$, the sequence (D_n) is an increasing sequence of stopping times (7.3 Corollary 1) converging to ∞. Define the stopping time D'_n as the restriction of V_j to the set $\{D_n < \infty\}$. Because $\bar{G}_n$ is contained in $B_j =]U_j, V_j]$ for every n, D'_n is a stopping time and we have $G_n \subseteq]D_n, D'_n]$ for every n.

According to 9.9 Corollary 2 the sequence $(X_{D'_n} - X_{D_n})$ converges to 0 almost surely as n tends to ∞ and since X is of class (D) this convergence takes place in the $L^1(P)$-norm.

Thus we finally get, taking into account (9.13.1):

$$\lim_n |\mu_X|(A_n \cap B_j) \le \lim_n |\mu_X|(G_n) + \varepsilon$$
$$\le \lim_n |\mu_X|(]D_n, D'_n]) + \varepsilon$$
$$\lim_n |\mu_X(]D_n, D'_n]) + \varepsilon_j + \varepsilon$$
$$= \varepsilon_j + \varepsilon .$$

After the preceding argument, that proves the σ-additivity of μ_X.

Examining the last part of the proof above, we get the following

<u>Corollary 1</u>: For a quasipotential $P = (P_t)$ the following assertions are equivalent:

a) The associated measure μ_P is σ-additive.

b) $P = (P_t)$ is of class (D).

c) $P = (P_t)$ is locally of class (D) i.e. satisfies the condition:

For every $t \in R_+$ the set of random variables $\{P_T : T$ is a stopping time bounded by $t\}$ is uniformly integrable.

For further reference we note the

<u>Corollary 2</u>: A quasimartingale $X = (X_t)$ of class (D) has a unique decomposition $X = M + V$ into a uniformly integrable martingale $M = (M_t)$, zero at time $t=0$ and a predictable process $V = (V_t)$ of integrable variation. If we define differently from our convention at the beginning of this section V_∞ as $\lim_{t \to \infty} V_t$ and X_∞ as $\lim_{t \to \infty} X_t$, then V is uniquely determined by the conditions $V_0 = X_0$ and $E(V_T - V_S) = E(X_T - X_S)$ for every pair of stopping times $S \leq T$.

<u>Proof</u>: The only fact that needs proving is that the equality $E(V_T - V_S) = E(X_T - X_S)$ holds for every pair of stopping times $S \leq T$. This however is obvious from the fact that every uniformly integrable right continuous martingale $M = (M_t)$ with $M_0 = 0$ satisfies $E(M_T) = E(M_0) = 0$ for every stopping time T by virtue of Doob's Optional Sampling Theorem (Theorem 5.5).

A right continuous process $X = (X_t)$ is called a *local martingale*, if there exists an increasing sequence (T_n) of stopping times T_n with $\lim_n T_n = \infty$, such that for every n the stopped process X^{T_n} is a (uniformly integrable) martingale.

This definition allows us to characterize the case where μ_X is purely finitely additive (i.e. orthogonal to every σ-additive measure on R).

<u>Corollary 3</u>: Let $X = (X_t)$ be a quasimartingale and suppose that μ_X is purely finitely additive. Then X is a local martingale.

<u>Proof</u>: Define the stopping times T_n by $T_n(\omega) = \inf\{t \in R_+ : X_t(\omega) \leq n\}$ and put $S_n = T_n \wedge n$. In view of Theorem 9.9 and 5.3 Corollary 1 the process (X_t) has P-almost surely left hand limits, hence $\lim_n T_n = \infty$ holds for the increasing sequence (T_n) of stopping times. Denote by $X^n = (X_t^n)$ the process $X = (X_t)$, stopped at T_n. We are going to prove that X^n is a uniformly integrable martingale for every n.

Decompose μ_{X^n} into $\mu_{X^n} = \mu_M + \mu_P$ as in the proof of the Riesz decomposition theorem (9.11), defining μ_M by $\mu_M(H) = \mu_{X^n}(H \cap [0])$ for $H \in R$. The process X^n satisfies $|X_t^n| \leq n + |X_{T_n}|$, hence X^n if of class (D) and therefore μ_{X^n} is σ-additive, which implies that the measures μ_M and μ_P are σ-additive. On the other hand we have $\mu_P(H) = \mu_X(H \cap]0, T_n])$ for every element $H \in R$ and hence $|\mu_P| \leq |\mu_X|$. Now μ_X is purely additive by hypothesis, consequently μ_P is at the same time purely additive and σ-additive and therefore μ_P is equal to 0.

Thus $\mu_{X^n} = \mu_M$ holds and since μ_{X^n} is σ-additive, X^n is a uniformly integrable martingale.

In order to prove the final decomposition theorem, we need the following

<u>9.14 Lemma</u>: Let $X = (X_t)$ be a quasimartingale such that μ_X is purely finitely additive. Then $\lim_{t \to \infty} X_t = 0$ holds P-almost surely.

<u>Proof</u>: In view of Proposition 9.8 and Theorem 9.9, together with the remark preceding Theorem 9.9, we may assume that X is a negative submartingale, i.e. that μ_X is positive. The limit $f = \lim_{t \to \infty} X_t$ exists according to

9.9 Corollary 2 and we have $-E(X_F\, f) \leq -\varliminf_{t\to\infty} E(X_F X_t)$ for every $F \in \bigcup_{t\in\mathbb{R}_+} F_t$, because X is a submartingale. If we denote by $M = (M_t)$ a right continuous version of the martingale $t \to E(f|F_t)$, then $0 \leq \mu_M \leq \mu_X$ holds for the σ-additive measure μ_M. Hence μ_M is at the same time purely additive and σ-additive and must therefore be equal to zero, which proves that $f = 0$ holds P-almost surely.

The decomposition theorem is:

9.15 Theorem: Let $X = (X_t)$ be a quasimartingale. X can be written as a sum $X = M + N + V + P$ where

$M = (M_t)$ is a uniformly integrable right continuous martingale,

$N = (N_t)$ is a martingale satisfying $\lim_{t\to\infty} N_t = 0$ P-almost surely,

$V = (V_t)$ is a predictable process of integrable variation, zero at time $t = 0$.

and $P = (P_t)$ is a quasipotential which is at the same time a local martingale.

This decomposition is unique up to indistinguishability.

Proof: In view of the uniqueness of the Riesz decomposition and the Doob-Meyer decomposition, the uniqueness of the decomposition above is easily proved.

In order to prove the existence, decompose μ_X into its σ-additive part μ_σ and its purely finitely additive part μ_f.

The quasimartingale Y corresponding to μ_σ splits into the sum $Y = M + V$ according to Theorem 9.12, and to the quasimartingale Z corresponding to μ_f we apply the Riesz decomposition (Theorem 9.11), which yields $Z = N + P$. That N and P have the required properties follows from 9.13 Corollary 3 and Lemma 9.14.

In the following Corollary we list some properties of the preceding decomposition, which are easily proved in view of the correspondence between quasimartingales and measures:

<u>Corollary</u>: The decomposition $X = M + N + V + P$ has the following properties:

a) $M = 0$ if and only if $\lim_{t \to \infty} X_t = 0$ holds almost surely.

b) $N = 0$ if and only if (X_t) converges in the $L^1(P)$-norm as t tends to ∞.

c) $N = P = 0$ if and only if X is of class (D).

d) $P = 0$ if and only if X is locally of class (D).

3 Stochastic integration

§10 INTEGRATION WITH RESPECT TO MEASURE WITH VALUES IN A BANACH SPACE

This section gives a survey of the properties of vector valued measures we shall need further on. Although stochastic integration can be done without mentioning explicitly vector valued measures at all, using this well established theory at least makes the issue clearer and provides more straightforward proofs for the existence and uniqueness of stochastic integrals.

The only problem that arises is, whether a Banach valued measure, defined and σ-additive on a algebra of sets A, has an extension to the σ-algebra, generated by A. We shall see that this is no serious problem at all in the case where the measure has its values in a Banach space E of the form

$$E = L^p(P) \quad (1 \leq p < \infty).$$

First let us fix a set S, an algebra A of subsets of S and let Σ be the σ-algebra, generated by A. Remember that every additive set function on A res. Σ is called measure. The Banach space of *bounded* real measures on A resp. Σ is denoted by $ba(A)$ resp. $ba(\Sigma)$, the subspaces of σ-additive bounded measures by $ca(\Sigma)$ resp. $ca(\Sigma)$. Every σ-additive real measure on a σ-algebra Σ is bounded and every bounded σ-additive real measure on an algebra A has a unique extension to a σ-additive measure on the σ-algebra Σ, generated by A.

Recall the following theorem on weak compactness of sets of real measures (see e.g. [7] Chap. IV):

<u>10.1 Theorem:</u> A subset $K \subseteq ca(\Sigma)$ of σ-additive real measures is weakly

sequentially compact if and only if K is bounded and the σ-additivity is uniform for elements $\mu \in K$.

Moreover, if K is weakly sequentially compact, there is a positive measure $\lambda \in ca(\Sigma)$, satisfying:

(a) $\lambda(A) \leq \sup_{\mu \in K} |\mu(A)|$ for every $A \in \Sigma$.

(b) The limit $\lim_{\lambda(A) \to o} \mu(A) = o$ is uniform with respect to $\mu \in K$.

<u>Remark 1</u>: (1) The weak topology on $ca(\Sigma)$ is the topology induced by the duality $\langle ca(\Sigma), B(\Sigma) \rangle$, where $B(\Sigma)$ is the space of bounded measurable functions on S.

(2) A subset K of a topological space E is called sequentially compact if every sequence of elements of K has a subsequence, converging in E.

Theorem 10.1 is essentially a consequence of the Vitali-Hahn-Saks Theorem ([7] Chap. III 7.2). The following is an easy consequence of this theorem:

<u>10.2 Proposition</u>: Let (μ_n) be a sequence of σ-additive measures $\mu_n \in ca(\Sigma)$. If $\mu(A) = \lim_n \mu_n(A)$ exists for every $A \in \Sigma$, then μ is countably additive and the σ-additivity is uniform for $n = 1, 2, \ldots$

Let now M be a *bounded* measure on A with values in a Banach space E. The *variation of* M is defined as $vM(A) = \sup \Sigma_i \|M(A_i)\|$ $(A \in A)$ and the *semi-variation of* M as $svM(A) = \sup \|\Sigma_i a_i M(A_i)\|$ $(A \in A)$, where the suprema are taken over all finite partitions (A_i) of A into disjoint sets $A_i \in A$, and all finite sequences (a_i) of real numbers a_i satisfying $|a_i| \leq 1$.

The variation vM of M is an additive (not necessarily finite) set function on A; if vM is finite M is said to be of *finite variation*.

In this case vM is a positive real measure on A which is σ-additive if and only if M is σ-additive.

For bounded real measures $\mu \in ba(A)$ variation and semivariation coincide, and we write $|\mu|$ instead of $v\mu$. In general we always have

$$o \leq svM(A) \leq vM(A)$$

and svM need not to be additive, however in contrast to vM the semivariation svM is always bounded. In fact if, for an element x' of the dual E' of the Banach space E, we denote by $\langle M, x' \rangle$ the real measure

$$A \to \langle M(A), x' \rangle ,$$

we get $svM(S) = \sup\{|\langle M,x'\rangle(S)|: x' \in E', \|x'\| \leq 1\}$

$$\leq \sup\{2 \sup_{A \in \mathcal{A}} |\langle M,x'\rangle(S)| : x' \in E', \|x'\| \leq 1\}$$

$$\leq 2 \sup_{A \in \mathcal{A}} \|M(A)\| .$$

It is convenient to extend the set function svM to the family of all subsets B of S, defining $svM(B) = \inf\{svM(A) : B \subseteq A \text{ and } A \in \mathcal{A}\}$. A subset $B \subseteq S$ is called M-*null set* if $svM(B) = o$ holds, the exceptional sets for M-almost everywhere convergnece etc. are these M-null sets.

For two real functions f and g on S we define

$$d(f,g) = \inf_{\alpha \geq o}(\alpha + svM(\{|f - g| \geq \alpha\})) .$$

d is a pseudo metric on the space $\mathbb{R}^S$ of real functions on S, and we have $d(f,o) = o$ if and only if $f = o$ M-almost everywhere. Convergence with respect to the topology, generated by d, is called *convergence in M-measure*.

10.3 Definition: Let M be a bounded measure on an algebra $\mathcal{A}$ of subsets of the set S with values in a Banach space E. The integral $\int f \, dM$ of a simple (real) function $f = \Sigma_i \alpha_i \chi_{A_i}$ ($\alpha_i \in \mathbb{R}, A_i \in \mathcal{A}$) is an element of E,

101

defined by $\int f \, dM = \Sigma_i a_i M(A_i)$.

A real function f on S is called *M-integrable* if there exists a sequence (f_n) of simple functions f_n, satisfying:

(a) (f_n) converges to f in *M*-measure.

(b) $\lim_{svM(A)\to 0} \int \chi_A f_n \, dM = 0$ $(A \in \mathcal{A})$ uniformly in $n = 1, 2, \ldots$

10.4 Proposition: Let f be an *M*-integrable function and (f_n) be a sequence of simple functions, satisfying (a) and (b) of Definition 10.3.

For every $A \in \mathcal{A}$ $\chi_A f$ is integrable and the sequence $(\int \chi_A f_n \, dM)$ converges to an element of E uniformly in $A \in \mathcal{A}$. The element $\lim_n \int f_n \, dM$ is called the integral of f with respect to M.

The set function $f \, dM$ on $\mathcal{A}$, which maps elements $A \in \mathcal{A}$ to $\int \chi_A f \, dM$ is an additive bounded measure on $\mathcal{A}$, satisfying $\lim_{svM(A)\to 0} sv(f \, dM)(A) = 0$.

Proof: For every $\varepsilon > 0$ there exists a $\delta > 0$ such that $svM(A) < \delta$ implies $\|\int \chi_A f_n \, dM\| < \varepsilon$ for all $n = 1, 2, \ldots$, according to 10.3 (b). 10.3 (a) yields an integer k such $m, n \geq k$ implies $svM(F) < \delta$, where F is the set $F = \{|f_m - f_n| \geq \varepsilon/svM(S)\}$.

If A is an arbitrary element of $\mathcal{A}$, we have

$$\|\int \chi_A f_n \, dM - \int \chi_A f_m \, dM\| = \|\int \chi_A (f_n - f_m) \, dM\|$$

$$\leq \|\int \chi_A \chi_F f_n \, dM\| + \|\int \chi_A \chi_F f_m \, dM\|$$

$$+ \|\int \chi_A (1 - \chi_F)(f_n - f_m) \, dM\|$$

$$< 3\varepsilon,$$

which proves that $\int \chi_A f_n \, dM$ converges uniformly in $A \in \mathcal{A}$. The second assertion is an obvious consequence of the first.

10.5: The vector space $S(\mathcal{A})$ (resp. $S(\Sigma)$) of simple measurable functions forms a subspace of the vector space of bounded real functions on S. By $B(A)$ (resp. $B(\Sigma)$) we denote the closure of $S(A)$ (resp. $S(\Sigma)$) in this space with respect to the supremum-norm

$$\|f\| = \sup \{f(s) : s \in S\}.$$

Clearly every $f \in B(A)$ is M-integrable, and the linear operator $T : B(\Sigma) \to E$, defined by $T(f) = \int f \, dM$, is an operator of norm $\|T\| = svM(S)$.

This yields a 1-1 correspondence between bounded additive measures on A with values in E, and continuous linear operators from $B(A)$ to E, which is moreover an isometric isomorphism for the obvious Banach space structures.

In future we shall therefore write M for the measure and for the associated operator T defined above, and having already abused notation, we are going to attach a third meaning to M.

If we define on $B(A)$ the seminorm

$$\|f\|_\infty = \inf\{a \in \mathbb{R}_+ : |f| \leq a, \ M\text{-almost everywhere}\},$$

the quotient $B^\infty(M) = B(A)/N$, where N is the subspace of all null functions in $B(A)$, is a Banach space under the quotient norm. Clearly N is the kernel of the linear operator M, and M can be considered as linear operator on $B^\infty(M)$.

Studying the adjoints M' and biadjoints M'' of the operator M yields useful information about the measure M, in case that M is σ-additive. First note that we identify the dual of $B(A)$ with the space $ba(A)$ of real bounded measures on A, i.e. we have:

$$M : B(A) \to E$$

and $M' : E' \to ba(A)$.

For an element $x' \in E'$ $M'(x')$ is the real measure $A \to \langle M(A), x' \rangle$, which was denoted by $\langle M, x' \rangle$.

If the measure M is weakly σ-additive on A, then M' maps E' into $ca(A)$. Now every bounded real σ-additive measure on A has a unique extension to a σ-additive measure on Σ, the σ-algebra generated by A, and identifying $ca(A)$ with $ca(\Sigma)$, we can replace the last diagram by
$$M' : E' \to ba(\Sigma),$$
and we get, restricting M'' to $B(\Sigma)$, which is a subspace of $ba(\Sigma)'$ under evaluation
$$M'' : B(\Sigma) \to E''.$$
Keeping in mind that the image of M' is contained in $ca(\Sigma)$, we get the following

<u>10.6 Proposition</u>: Every weakly σ-additive bounded measure M on A with with values in E has a unique extension to a bounded $\sigma(E'',E')$-σ-additive measure on Σ with values in the bidual E'' of E.

As we shall see, weak σ-additivity implies (norm) σ-additivity, so that the problem of extending σ-additive measures on A to Σ reduces to the question, whether M'' maps $B(\Sigma)$ into E, which is a subspace of E'' under evaluation. This question can be positively answered if E is reflexive, or more generally, if M is a weakly compact map. However, before studying this problem more deeply, we first look at the vectorspace of M-integrable functions in the case where M is a σ-additive measure on a σ-algebra Σ.

<u>10.7 Proposition</u>: If M is a countably additive measure on a σ-algebra Σ with values in a Banach space E, the set of real measures $\{\langle M,x'\rangle : x' \in E', \|x'\| \leq 1\}$ is weakly sequentially compact in $ca(\Sigma)$.

<u>Proof</u>: We have $|\langle M,x'\rangle(A)| \leq |\langle M,x\rangle|(S) < \infty$ for every $x' \in E'$ because M is countably additive, therefore the set $\{M(A) : A \in \Sigma\}$ is weakly bounded in E and hence norm bounded. That the countable additivity of the family $\{\langle M,x'\rangle : x' \in E', \|x'\| \leq 1\} \subseteq ca(\Sigma)$ is uniform, is immediate from

the countable additivity of M. That completes the proof by virtue of Theorem 10.1.

In view of Theorem 10.1 we get the following corollary:

<u>Corollary</u>: Let be a σ-additive measure on Σ. There exists a positive countably additive measure λ, defined on Σ, satisfying:

(a) $\lambda(A) \leq svM(A)$ $(A \in \Sigma)$

(b) $\lim_{\lambda(A) \to o} svM(A) = o$.

We denote by $L^1(M)$ the vector space of (equivalence classes of M-almost everywhere coinciding) M-integrable real functions.

The following theorem characterizes $L^1(M)$ as the completion of the space of simple measurable functions:

<u>10.8 Theorem</u>: Let S be the vector space of Σ-measurable simple functions and let M be a σ-additive measure with values in the Banach space E, defined on the σ-algebra Σ. Define the norm $\|f\| = sv(f\,dM)(S)$ on S, where fdM is the E-valued measure given by

$$A \to M(\chi_A f) \quad (A \in \Sigma).$$

We have:

(a) Every Cauchy sequence (f_n) of elements of S converges in M-measure to a function f.

(b) The space $L^1(M)$ of M-integrable (real) functions can be identified with the completion of S with respect to the norm, given above.

(c) The linear operator $M : L^1(M) \to E$, given by $M(f) = \int f\,dM$, is continuous with norm $\|M\| \leq 1$.

Proof: (a) Let g be an element of S, and $a > 0$. We have

$$\|g\| = \sup \{\|M(gh)\| : |h| \leq 1, \ h \in S\}$$
$$= \sup \{\|M(h)\| : |h| \leq |g|, \ h \in S\}$$
$$\geq \sup \{\|M(h)\| : |h| \leq a \ \chi_{\{|g| \geq a\}}, \ h \in S\}$$
$$= a \sup \{\|M(h)\| : |h| \leq \chi_{\{|g| \geq a\}}, \ h \in S\}$$
$$= a \ svM(\{|g| \geq a\}).$$

Consequently the metric d of convergence in M-measure satisfies

$$d(g,o) \leq \inf_{a>0}(a + 1/a\|g\|) = (\|g\|)^{1/2}.$$

Thus the Cauchy sequence (f_n) of elements of the normed space S is a Cauchy sequence for the metric of convergence in M-measure and hence (f_n) is a Cauchy sequence for the metric of convergence in λ-measure, where λ is a positive σ-additive measure, satisfying (a) and (b) of 10.7 Corollary. (f_n) converges therefore to a function f in λ-measure (see [7] III.6.5). This completes the proof of (a), because in view of 10.7 Corollary (b) convergence in λ-measure implies convergence in M-measure.

(b) Let (f_n) be a Cauchy sequence in S. By (a) (f_n) converges to a function f in M-measure, i.e. (f_n) and f satisfy condition (a) of Definition 10.3, and we have to show that 10.3 (b) is satisfied too. Let an $\varepsilon > 0$ be given. There is an m, such that $\|f_n - f_m\| \leq \varepsilon$ holds for every $n \geq m$, and we have

$$\lim_{svM(A) \to 0} [\sup_{n \geq 1} \|M(\chi_A f_n)\|] = \lim_{svM(A) \to 0} [\sup_{n \geq m} \|M(\chi_A f_n)\|]$$

$$\leq \lim_{svM(A) \to 0} [\|M(\chi_A f_m)\| + \sup_{n \geq m} \|f_n - f_m\|]$$

$$\leq \varepsilon.$$

According to Definition 10.3, this shows that f is an M-integrable function with (f_n) satisfying 10.3 (a) and (b). Let on the other hand (f_n) be a sequence of simple functions in S, satisfying (a) and (b) of 10.3. For a given $\varepsilon > 0$ there exists an m such that $\|M(\chi_A(f_n-f_m))\| \leq \varepsilon$ holds for all $A \in \Sigma$ and $n \geq m$, according to Proposition 10.4. Hence

$$\|f_n - f_m\| = sv[(f_n - f_m)dM](S) \leq 2 \sup_{A \in \Sigma} \|M(\chi_A(f_n - f_m))\| \leq 2\varepsilon$$

which proves that (f_n) is a Cauchy sequence in S. That the assertion (c) holds is evident.

<u>Corollary 1</u>: Let f be an element of $L^1(M)$. The element $M(f) \in E$ is uniquely determined by $\langle M(f), x' \rangle = \int f\, d\langle M, x' \rangle$, $(x' \in E')$.

<u>Proof</u>: Clearly the equality above uniquely determines $M(f)$, provided it all holds for all $x' \in E'$. For simple functions $g \in S$ the equality holds by definition, and we have

$$\int |g|\, d|\langle M, x' \rangle| = \sup \{|\int gh\, d\langle M, x' \rangle| : h \in S,\ |h| \leq 1\}$$
$$= \sup \{\langle M(gh), x' \rangle : h \in S,\ |h| \leq 1\}$$
$$\leq \|x'\|\ \|g\|$$

for every $x' \in E'$.

Let now f be an element of $L^1(M)$ and (f_n) a Cauchy sequence in $L^1(M)$ of elements $f_n \in S$, converging to f.

According to Proposition 10.4 we have $\langle M(f), x' \rangle = \lim_n \langle M(f_n), x' \rangle$ for every $x' \in E'$. On the other hand, in view of the argument above, (f_n) is a Cauchy sequence in $L^1(|\langle M, x' \rangle|)$ for $x' \in E'$, hence

$$\lim_n \langle M(f_n), x' \rangle = \lim_n \int f_n\, d\langle M, x' \rangle = \int f\, d\langle M, x' \rangle$$

holds for all

$x' \in E'$, i.e. we have $\langle M(f), x' \rangle = \int f\, d\langle M, x' \rangle$ for all $x' \in E'$.

Corollary 2: A Σ-measurable real function f is M-integrable if and only if for every $x' \in E'$ f is $\langle M, x' \rangle$-integrable and the family

$$\{f d\langle M, x' \rangle : \|x'\| \leq 1, \; x' \in E'\}$$

is weakly sequentially compact.

Proof: That the condition is necessary is obvious in view of Proposition 10.4, let us prove the sufficiency.

For every positive integer n the function $f_n = f \chi_{A_n}$, where A_n is the set $A_n = \{|f| \leq n\}$ is M-integrable and we have

$$\lim_{n \to \infty} |f \, d\langle M, x' \rangle| (A_n^c) = 0$$

for every $x' \in E'$. Because of the weak sequential compactness this convergence takes place uniformly in $x' \in E'$ with $\|x'\| \leq 1$, i.e.

$$\lim_{n \to \infty} [\sup_{\|x'\| \leq 1} |f \, d\langle M, x' \rangle| (A_n^c)] = 0$$

and consequently

$$\lim_{n \to \infty} [\sup_{\|x'\| \leq 1} |(f - f_n) d\langle M, x' \rangle| (S)] = 0$$

holds. This shows that (f_n) is a Cauchy sequence in $L^1(M)$ converging to f, which completes the proof.

As an immediate consequence we have

Corollary 3: (Lebesgue dominated convergence theorem).
Let (f_n) be a sequence of M-integrable functions converging to a function f in M-measure and satisfying $|f_n| \leq |g|$ for a M-integrable function g and all n.

Then f is M-integrable and the sequence (f_n) converges to f in the $L^1(M)$-norm.

We conclude this section by studying the 'extension problem' for a bounded σ-additive measure, defined on an algebra of sets A.

10.9 Lemma: Let A be an algebra of subsets of the positive integers N, containing all sets of the form $\{n\}$ $(n \in N)$. Every bounded weakly σ-additive measure M on A with values in a weakly sequentially complete subset of a Banach space E has a unique norm-σ-additive extension to a measure with values in E, defined on the σ-algebra Σ, generated by A.

Proof: According to Proposition 10.6, M has a $\sigma(E'', E')$-σ-additive extension $\bar{M}$ to Σ with values in E''. Now every element of Σ is a countable disjoint union of elements of A and, because M has its values in a weakly sequentially complete subset of E, the extension $\bar{M}$ has its values in E. Furthermore we may suppose that E is separable or more precisely that the set $\{M(\{n\}) : n \in N\}$ is total in E, and what remains to prove is that every weakly σ-additive measure $\bar{M}$, defined on Σ with values in E, is norm-σ-additive. Suppose there is a sequence (A_k) of elements $A_k \in \Sigma$, such that $\cap_k A_k = \phi$ and $\|\bar{M}(A_k)\| \geq \varepsilon$ holds for all k and an $\varepsilon > 0$. There exists a sequence (x'_k) of elements of E', satisfying $\|x'_k\| = 1$ and $\langle \bar{M}, x'_k \rangle (A_k) = \|\bar{M}(A_k)\|$. Using a Cantor diagonal procedure, there is a subsequence $(y'_m = x'_{k_m})$ of (x'_k) such that $\lim_m \langle x, y'_m \rangle$ exists for every element $x \in \{M(\{n\}) : n \in N\}$ hence $\lim_m \langle x, y'_m \rangle$ exists for every $x \in E$ by virtue of the Banach-Steinhaus theorem ([7]chap. II, 3.6). In particular $\lim_m \langle \bar{M}, y'_m \rangle (A)$ exists for every element $A \in \Sigma$, hence, according to Proposition 10.2, the sequence of real measures $(\langle \bar{M}, y'_m \rangle)$ is uniformly σ-additve, which is absurd.

10.10 Theorem: Let M be a weakly σ-additive measure on an algebra A with values in a Banach space E.

The following assertions are equivalent:

(a) M has a unique extension to a (norm) σ-additive measure on the σ-algebra Σ, generated by A, with values in E.

(b) M maps A into a relatively weakly compact subset of E.

(c) M maps A into a weakly sequentially complete bounded subset of E.

(d) M has an extension to a weakly σ-additive measure on the Σ-algebra, generated by A.

Proof: (a) $\Rightarrow$ (b): We have to show that every σ-additive measure $\bar{M}$ defined on a σ-algebra Σ, maps Σ to a relative weakly compact subset of E. Firstly we observe that $\bar{M}$ is weakly bounded and hence norm bounded. Let λ be a positive $\bar{M}$-absolutely continuous measure as in 10.7 Corollary. $B^\infty(M)$ can be identified with $L^\infty(\lambda)$, the map $f \to \bar{M}(f)$ from $B^\infty(\bar{M})$ to E (see 10.5) can therefore be considered as continuous linear operator from $L^\infty(\lambda)$ to E. Since $\bar{M}$ is weakly σ-additive, the adjoint $\bar{M}'$ of this operator maps E' into $L^1(\lambda)$ (use the Radon-Nikodym theorem), hence the operator $\bar{M} : L^\infty(\lambda) \to E$ is continuous for the topologies $\sigma(L^\infty(\lambda), L^1(\lambda))$ and $\sigma(E, E')$. Now the unit ball of $L^\infty(\lambda)$ is $\sigma(L^\infty(\lambda), L^1(\lambda))$ compact, consequently the operator $\bar{M}$ is a weakly compact operator, which completes the proof of the implication (a) $\Rightarrow$ (b).

(b) $\Rightarrow$ (c): It is clear that any weakly compact set in E is weakly sequentially complete and bounded.

(c) $\Rightarrow$ (d): According to Proposition 10.6 M has an extension $\bar{M}$ to a $\sigma(E'', E')$-countably additive measure on Σ with values in E'', we have to prove that $\bar{M}$ actually takes its values in E. Let A_d be the smallest family of elements of Σ, which contains A and is stable under countable intersections. Every $B \in A_d$ can be written as intersection of a decreasing sequence of elements of A, assertion (c) therefore implies that $\bar{M}(B)$ is an element of E for every $B \in A_d$.

Now for every element $A \in \Sigma$, the net $\{\bar{M}(B) : B \in A_d, B \subseteq A\}$ is a weak Cauchy-net in E, and we shall show that this net is in fact a Cauchy-net for the norm on E, which will complete the proof of the implication (c) $\Rightarrow$ (d).

Assume that there is an increasing sequence (B_n) of elements of A_d with $B_n \subseteq A$ for all n, such that $\|\bar{M}(B_n) - \bar{M}(B_{n+1})\| \geq \varepsilon$ holds for an $\varepsilon > 0$ and all n. If we restrict ourselves to the algebra generated by the sequence (B_n), everything reduces to the case of Lemma 10.9, which shows that our assumption was absurd. (d) $\Rightarrow$ (a): For the proof of this assertion the last part of the proof of the assertion (c) $\Rightarrow$ (d) can be used.

Corollary 1: Every weakly σ-additive measure with values in a Banach space E, defined on a σ-algebra Σ, is bounded and (norm) σ-additive.

Corollary 2: A measure M, defined on an algebra A, with values in a weakly sequentially complete Banach space E has an extension to a σ-additive measure on the σ-algebra Σ, generated by A, if and only if M is weakly σ-additive and bounded, i.e. if and only if for every $x' \in E'$ the real measure $\langle M, x' \rangle$ on A is σ-additive and bounded.

§11 STOCHASTIC INTEGRATION WITH RESPECT TO SQUARE INTEGRABLE MARTINGALES AND MARTINGALES OF BOUNDED VARIATION.

In this section the stochastic integral with respect to a right continuous stochastic process is defined. This leads to the notion of a summable process, i.e. a stochastic process with respect to which a stochastic integral can be defined. A necessary and sufficient condition for a stochastic process to be summable is given; this condition hints at the importance of quasimartingales as a tool for stochastic integration. The classical example of stochastic integrals with respect to square integrable martingales is used to illustrate the definitions. For this example the equivalence of our definition and the definition of the stochastic integral as it can be found, e.g. in [5] is explicitly established. Spaces of square integrable martingales which are stable under stopping are studied in the final part of this section.

As usual, we fix a stochastic base $(\Omega, F, P, (F_t), \bar{R}_+$. Stochastic processes, defined only for $t \in R_+$ will always have the property that $\lim_{t \to \infty} X_t$ exists P-almost surely and X_∞ is defined in this case as $X_\infty = \lim_{t \to \infty} X_t$.

11.1 Definition: Denote by A the Boolean algebra of subsets of $R_+ \times \Omega$, generated by stochastic intervals of the form $[0_F]$ $(F \in F_0)$ and $]S,T]$, $S \le T$, S,T stopping time. (Recall that 0_F is the stopping time which is equal to 0 on F and equal to ∞ on F^c.) The σ-algebra generated by A is the σ-algebra of predictable sets (see 7.4).

Let $X = (X_t)$ be a right continuous adapted stochastic process which is at the same time a $L^p(P)$-process (see 1.3) for a p $(1 \le p < \infty)$.

X generates an additive measure I_X with values in $L^p(P)$ on A, which is defined by

$$I_X(A) = \chi_F X_{o_F} + \sum_{i=1}^{n} X_{T_i} - X_{S_i}$$

where

$$A = [o_F] \cup]S_1, T_1] \cup \ldots \cup]S_n, T_n] \quad (F \in F_o, \ S_i < T_i$$

on $\{T_i < \infty\}$ and $T_i < S_{i+1}$ on $\{S_{i+1} < \infty\}$) is a typical element of A.

If the measure I_X admits a σ-additive extension to a measure on Σ_p the σ-algebra of predictable sets, with values in $L^p(P)$ $(1 \le p < \infty)$, the process $X = (X_t)$ is called p-*summable* or simply *summable* if $p = 1$.

For a p-summable process X we call the σ-additive extension of I_X to Σ_p the *stochastic measure generated by* X. The stochastic measure generated by X will again be denoted by I_X.

Since every Banach space $L^p(P)$ $(1 \le p < \infty)$ is weakly sequentially complete, I_X has a σ-additive extension to Σ_p if and only if I_X is weakly σ-additive and bounded on A (see 10.10 Corollary 2).

We met an example of a summable process already in §2, namely the adapted process of integrable variation $V = (V_t)$. It is clear that the mapping $Z \to I_V(Z)$, defined in 2.7, yields the stochastic measure generated by V if we restrict I_V to Σ_p-measurable stochastic processes Z. This stochastic measure is moreover of bounded variation, the variation vI_V being equal to the variation μ_V of the real measure μ_V generated by V (see §2).

The following theorem provides the argument we shall use in the sequel to check whether a stochastic process $X = (X_t)$ is summable.

<u>11.2 Theorem</u>: Let $X = (X_t)$ be a right continuous adapted stochastic process p a real number $1 \le p < \infty$ and $q = p/p - 1$ $(q = \infty$ for $p = 1)$.

For every $f \in L^q(P)$ denote by $F = (F_t)$ a right continuous modification of the martingale $t \to E(f, F_t)$. The stochastic process $X = (X_t)$ is p-summable if and only if the process $XF = (X_t F_t)$ is a quasimartingale of class (D) for every $f \in L^q(P)$.

113

Proof: Let the measure I_X on A be defined as in 11.1. According to the definition and 10.10 Corollary 2 X is p-summable if and only if for every $f \in L^q(P)$ the real measure $\langle I_X, f \rangle$ (see 10.5) is σ-additive and bounded on A.

Now for every $f \in L^q(P)$ we have
$$\langle I_X, f \rangle([o_F]) = E(X_F X_o f) = E[X_F X_o E(f|F_o)]$$
and
$$\langle I_X, f \rangle(]S,T]) = E[(X_T - X_S)f] = E[X_T E(f|F_T) - X_S E(f|F_S)].$$
Since F_T equals $E(f|F_T)$ for every stopping time T according to the optional sampling theorem for martingales (Theorem 5.5), the real measure $\langle I_X, f \rangle$ is bounded on A if and only if the process $XF = (X_t F_t)$ is a quasimartingale by virtue of the equivalence (a) $\leftrightarrow$ (b) of Theorem 9.2.

If $\langle I_X, f \rangle$ is σ-additive, then $\langle I_X, f \rangle$ has a unique extension to a σ-additive (real) measure on Σ_p. According to 8.8 there is a predictable process $V = (V_t)$ generating $\langle I_X, f \rangle$, i.e. satisfying $V_o = X_o F_o$ and $E(V_T - V_S) = E(X_T F_T - X_S F_S) = \langle I_X, f \rangle(]S,T])$ for any pair of stopping times $S \leq T$. It is clear that $XF - V = (X_t F_t - V_t)$ is a uniformly integrable martingale, consequently XF is a quasimartingale of class (D).

If on the other hand XF is a quasimartingale of class (D), XF can be written as a sum of a uniformly integrable martingale $M = (M_t)$ which is zero at time $t=o$ and a process of integrable variation $V = (V_t)$ according to 9.13 Corollary 2. We have $V_o = X_o F_o$ and $E(V_T - V_S) = E(V_T + M_T - V_S - M_S) = \langle I_X, f \rangle(]S,T])$ for every pair of stopping times $S \leq T$. Hence $\langle I_X, f \rangle$ is the restriction to A of the real measure generated by V and consequently σ-additive.

11.3: Let $X = (X_t)$ be a p-summable process, I_X the stochastic measure generated by X. Recall that for an I_X-integrable process

$$Z = (Z_t) \in L^1(I_X)$$

we write $I_X(Z)$ for the integral of Z with respect to I_X.

The integral operator I_X defines a bounded linear operator (see 10.5)

$$I_X : B^\infty(I_X) \to L^p(P) ,$$

the adjoint of which is given by $I'_X(f) = \langle I_X, f \rangle$ $(f \in L^q(P))$, or in other words $I'_X(f)$ is the countably additive (I_X-absolutely continuous) real measure $A \to E(I_X(\chi_A)f)$ $(A \in \Sigma_p)$. Since every evanescent subset of $R_+ \times \Omega$ is a I_X-null set, $\langle I_X, f \rangle$ is a stochastic measure $\langle I_X, f \rangle \in sca(\Sigma_p)$ for every $f \in L^q(P)$. According to 8.8 there is a unique predictable process of integrable variation $V^f = (V_t) \in pSV$ generating $\langle I_X, f \rangle$, i.e. satisfying $E(\int Z_s dV_s^f) = \int Z d\langle I_X, f \rangle$ for every predictable bounded process $Z = (Z_s)$. To make our notation consistent with the usual one, we write $F = (F_t)$ for the right continuous modification of the martingale

$$t \to E(F|F_t)$$

and we write $\langle X, F \rangle = (\langle X, F \rangle_t)$ for the process $V^f = (V_t^f)$. Checking the proof of Theorem 11.2 we get the following characterization of $\langle X, F \rangle$:

Let $X = (X_t)$ be a p-summable proces and f an element of $L^q(P)$. Denote by $F = (F_t)$ the right continuous modification of the martingale $t \to E(f|F_t)$. Then $XF = (X_t F_t)$ is a quasimartingale of class (D) and $\langle X, F \rangle = (\langle X, F \rangle_t)$ is the unique predictable process of integrable variation such that $XF - \langle X, F \rangle$ is a uniformly integrable martingale which is zero at time $t = 0$. $\langle X, F \rangle$ and $\langle I_X, f \rangle = I'_X(f)$ are related to each other according to the equality $E(\int Z_s d\langle X, F \rangle_s) = \int Z\, d\langle I_X(Z, f) \rangle$ which holds for every bounded predictable process $Z = (Z_s)$.

The family (Y_t) of elements of $L^p(P)$ $Y_t = I_X(\chi_{[0,t]} Z)$ $(Z \in L^1(I_X))$ forms a right continuous $L^p(P)$-process and, provided that this $L^p(P)$-process has a right continuous modification $X(Z) = (X(Z)_t)$, this modification is a

p-summable process, uniquely determined by

$$E(\langle X(Z), F\rangle_t) = E(\int_{[0,t]} Z_s d\langle X, F\rangle_s) \qquad (f \in L^q(P)).$$

In the case where X is a martingale the existence of the right continuous modification $X(Z)$ is no problem, because the $L^p(P)$-process (Y_t) is a martingale. Since quasimartingales were not defined as $L^p(P)$-processes, we have to be more careful in the general case.

<u>11.4 Theorem:</u> Let $X = (X_t)$ be a p-summable process, and $Z = (Z_t) \in L^1(I_X)$. There exists a right continuous modification $X(Z) = (X(Z)_t)$ of the $L^p(P)$-process $t \to I_X(\chi_{[0,t]} Z)$. $X(Z)$ is a p-summable process and is uniquely determined by

$$E(X(Z)_t f) = E(\langle X(Z), F\rangle_t) = E(\int_{[0,t]} Z_s d\langle X, F\rangle_s) \qquad (f \in L^q(P)).$$

<u>Proof:</u> After the argument of 11.3 we have only to prove the existence of the right continuous modification of the $L^p(P)$-process $t \to I_X(\chi_{[0,t]} Z)$. We shall do that by defining a measure μ, corresponding to $(I_X(\chi_{[0,t]} Z))$ in the sense to §9. Recall that R is the ring, generated by bounded stochastic intervals of the form $[o_F]$ $(F \in \bigcup_{t \in R_+} F_t)$ and $]S,T]$ (see 9.6).

We define the real measure μ on R by

$$\mu([o_F]) = -E[\chi_F I_X(Z)]$$

and $\mu(]S,T]) = E[I_X(\chi_{]S,T]} Z)] = \int \chi_{]S,T]} Z \, d\langle I_X, 1\rangle$

It is clear that μ satisfies (d) and (e) of condition (S) (see 9.6).

Let t be an element of R_+ and $F \in F_t$, then

$$|\mu|(]t_F, \infty]) = |Z d\langle I_X, 1\rangle|(]t_F, \infty]) \leq E(\chi_F V_\infty)$$

holds, i.e. μ satisfies condition (S). $V = (V_t)$ is the predictable process generating the real measure $|Z d\langle I_X, 1\rangle|$. According to 9.9 Corollary 1 there is a unique quasimartingale $Y = (Y_t)$, such that the associated measure μ_Y coincides with μ. Y is the desired right continuous modification of $t \to I_X(\chi_{[0,t]}Z)$.

11.5 Definition: Let $X = (X_t)$ be a p-summable process, and $Z = (Z_t)$ integrable with respect to I_X.

The right continuous modification $X(Z) = (X(Z)_t)$ of the $L^p(P)$-process $t \to I_X(\chi_{[0,t]}Z)$ is called the *stochastic integral of* Z *with respect to* X.

Let us now study an example of a 2-summable process. Let $M = (M_t)$ be a square integrable right continuous martingale, i.e. a right continuous martingale satisfying $\sup_t \|M_t\|_2 < \infty$. For every element $f \in L^2(P)$ the right continuous modification of $t \to E(f|F_t)$ is a square integrable martingale as well. Consequently, according to example (1) (following Definition 9.1) and the remark following Proposition 9.13, the process $MF = (M_t F_t)$ is a quasimartingale of class (D) for every $f \in L^2(P)$. By virtue of Theorem 11.2 the martingale $M = (M_t)$ is therefore a 2-summable process.

11.6 Theorem: Every right continuous square integrable martingale $M = (M_t)$ is a 2-summable process and the stochastic measure I_M, generated by M, has the following properties:

(a) The semivariation of the measure I_M is equal to
$$svI_M(A) = [E(\int \chi_A d\langle M,M \rangle_s)]^{1/2} = [\langle I_M, M_\infty \rangle(A)]^{1/2} \quad (A \in \Sigma_p)$$
and in particular
$$svI_M(\mathbb{R}_+ \times \Omega) = E(\langle M,M \rangle_\infty)^{1/2} = \|M_\infty\|_2.$$

(b) The space $L^1(I_M)$ can be identified with the Hilbert space $L^2(\langle I_M, M_\infty \rangle)$, in particular a predictable process $Z = (Z_t)$ is I_M-integrable if and only if $E(\int Z_s^2 d\langle M,M \rangle_s) < \infty$ holds.

(c) The integral operator $I_M : L^1(I_M) \to L^2(P)$ is an isometry from $L^1(I_M) = L^2(\langle I_M, M_\infty \rangle)$ to $L^2(P)$.

<u>Proof</u>: Let $Z = (Z_t)$ be an element of $L^1(I_M)$. The $L^1(I_M)$-norm of Z is (see 10.8):

$$\|Z\| = \sup\{\|I_M(Y)\|_2 : Y \in L^1(I_M), |Y| \leq |Z|\}$$

$$= \sup\{[E[(I_M(Y))^2]]^{1/2} : Y \in L^1(I_M), |Y| \leq |Z|\}.$$

In view of 11.3 (or Theorem 11.4) we have

$$E[(I_M(Y))^2] = E(\int Y_s^2 d\langle M, M\rangle_s) = \int Y^2 d\langle I_M, M_\infty\rangle,$$

and consequently

$$\|Z\|^2 = E(\int Z_s^2 d\langle M, M\rangle_s) = \int Z^2 d\langle I_M, M_\infty\rangle = E[(I_M(Z))^2].$$

Assertions (a), (b) and (c) are now an easy consequence of this equality.

The following Corollary follows immediately from Theorem 11.4, and the assertion, stated in the Corollary, serves as definition of the stochastic integral with respect to a square integrable martingale in [5].

<u>Corollary</u>: Let $M = (M_t)$ be a right continuous square integrable martingale and $Z = (Z_t) \in L^1(I_M)$, i.e. let Z be a predictable process, satisfying $E(\int Z_s^2 d\langle M, M\rangle_s) < \infty$. The stochastic integral of Z with respect to M is the unique right continuous square integrable martingale $M(Z) = (M(Z)_t)$, satisfying $E(M(Z)_\infty F_\infty) = E(\int Z_s d\langle M, F\rangle_s)$ for every right continuous square integrable martingale $F = (F_t)$.

<u>11.7</u> Theorem 11.6 (c) shows, that the image of $L^1(I_M)$ under I_M is a closed linear subspace of $L^2(P)$. An element $f \in L^2(P)$ is orthogonal to the range of I_M if and only if $E(I_M(Z)f) = 0$ holds for every $Z \in L^1(I_M)$

or equivalently, if and only if $\int Z d\langle I_M, f\rangle = E[\int Z_s d\langle M, F\rangle_s] = 0$ holds for all $Z \in L^1(I_M)$. Since $L^1(I_M)$ contains all bounded predictable processes, an element $f \in L^2(P)$ is therefore orthogonal to the range of I_M if and only if $\langle M, F\rangle = 0$, i.e. if and only if $MF = (M_t F_t)$ is a martingale, zero at time $t = 0$ (see 11.3).

For a stopping time T we have $I_M(\chi_{[0,T]}) = M_T$, and since the family of processes of the form $\chi_{[0,T]}$ is total in $L^1(I_M)$, the range of I_M is the closed subspace of $L^2(P)$, generated by the set $\{M_T : T$ is a stopping time$\}$.

More generally, a closed subspace $H \in L^2(P)$ is called *stable under stopping*, if and only if one of the three following equivalent conditions is satisfied:

(a) If f is an element of H and $F = (F_t)$ a right continuous modification of the martingale $t \to E(f|F_t)$, then F_T is an element of H for every stopping time T.

(b) If f is an element of H and $F = (F_t)$ a right continuous modification of the martingale $t \to E(f|F_t)$, then $F(Z)$ is an element of H for every process $Z \in L^1(I_F)$.

(c) If g is an element of $H^\perp$ then $\langle F, G\rangle = 0$ holds for every $f \in H$, where $F = (F_t)$ resp. $G = (G_t)$ are the right continuous modifications of the martingales $t \to E(f|F_t)$ resp. $t \to E(g|F_t)$.

That the conditions (a) - (c) are equivalent is obvious after the argument given before the definition. Notice that (c) implies that the orthogonal complement $H^\perp$ of a stable subspace $H \subseteq L^2(P)$ is stable under stopping and that $H^{\perp\perp} = H$.

An example of a subspace, stable under stopping, is the space H_c^2 of elements $f \in L^2(P)$, such that the right continuous modification $F = (F_t)$ of the martingale $t \to E(f|F_t)$ is a continuous martingale, zero at time $t = 0$.

Since for every stopping time T the stopped martingale $F^T = (F_{T \wedge t})$ is the right continuous modification of $t \to E(F_T|F_t)$, H_c^2 satisfies condition (a), and H_c^2 is closed in view of Doob's inequality (Theorem 4.11).

The orthogonal complement of H_c^2 is denoted by H_d^2, and, if a martingale $F = (F_t)$ is a right continuous modification of $t \to E(f|F_t)$ for an element $f \in H_d^2$, $F = (F_t)$ is called a *purely discontinuous martingale*. Note however, that in the case where F is a purely discontinuous martingale of bounded variation the paths of F are not necessarily purely discontinuous.

We now study the subspace H_d^2. Let T be a stopping time. Define H_T^2 to be the space of all $f \in H_d^2$ such that the corresponding martingale $F = (F_t)$ is continuous in the complement of the graph of T, or in other words, such that the set $\{(t, \omega) \in R_+ \times \Omega : F_t(\omega) \neq F_{t-}(\omega)\}$ is (up to an evanescent set) contained in the graph of T.

11.8 <u>Proposition</u>: Let T be a totally inaccessible or predictable stopping time and f an element of $L^2(P)$.

Write $F = (F_t)$ for the right continuous modification of the martingale $t \to E(f|F_t)$, $\Delta F_T = F_T - F_{T-}$, $A = \{T > 0\}$, T_A for the stopping time T restricted to A and $V = (V_t)$ for the dual predictable projection of the process $\Delta F_T \chi_{[T_A, \infty[}$.

The orthogonal projection of f onto H_T^2 is the element $g = \Delta F_T - V_\infty$.

<u>Remark</u>: The process $V = (V_t)$ is called the *compensating process* of

$\Delta F_T \chi_{[T_A, \infty[}$. If T is equal to zero almost surely, we have $g = F_0 = E(f|F_0)$; more generally, if T is predictable the process V is zero. If T is totally inaccessible, the process V is continuous. (See 8.10 Corollary 4).

Proof: According to 8.10 Corollary 4 $\Delta F_T \chi_{[T_A, \infty[} - V$ is a martingale and therefore the right continuous modification of $t \to E(g|F_t)$. By virtue of 8.10 Corollary 3 we get for every element $h \in L^\infty(P)$ (11.8.1):

$$E(gh) = E(\Sigma_s \Delta H_s \Delta G_s) = E(\Delta F_T \Delta H_T) \leq 2\|f\|_2 \|H_T\|_2$$

where $G = (G_s)$ (resp. $H = (H_s)$) are right continuous modifications of $s \to E(g|F_s)$ (resp. $s \to E(h|F_s)$).

Consequently $G = (G_s)$ is a square integrable martingale and hence (11.8.1) holds for all $h \in L^2(P)$. In particular g is an element of H_T^2 and in view of the remarks, preceding this proof, we have

$$E[(f - g)h] = E[(\Delta F_T - \Delta F_T)\Delta H_T] = 0$$

for every element h of H_T^2, which completes the proof.

By Zorn's lemma there is a maximal family of predictable resp. totally inaccessible stopping times with disjoint graphs, and we have the following

Corollary 1: Let $\{T_\alpha\}$ be a maximal family of predictable resp. totally inaccessible stopping times with disjoint graphs. The orthogonal projection of an element $f \in L^2(P)$ onto H_d^2 is given by $g = \Sigma_\alpha g_\alpha$, where g_α is the orthogonal projection of f onto $H_{T_\alpha}^2$.

In particular for an element $g \in H_d^2$ $E(g|F_t)$ is the orthogonal sum

$$E(g|F_t) = \Sigma_\alpha (\Delta G_{T_\alpha} \chi_{\{T_\alpha \leq t\}} - V_t^\alpha),$$

where $G = (G_t)$ is the right continuous modification of $t \to E(g|F_t)$, V^α

is the compensating process of $\Delta G_S X_{[S_\alpha, \infty[}$ and S_α the stopping time T_α, restricted to $\{T_\alpha > 0\}$.

<u>Corollary 2</u>: Let $F = (F_t)$ and $G = (G_t)$ be two square integrable martingales, G moreover purely discontinuous, i.e. G_∞ an element of H_d^2. Then the process $W = (W_t)$, defined by $W_t(\omega) = \Sigma_{s \leq t} \Delta F_s(\omega) \Delta G_s(\omega)$ is of integrable variation and $(F_t G_t - W_t)$ is a martingale zero at time $t = 0$.

<u>Proof</u>: Write $f = F_\infty$ and $g = G_\infty$, and let us first suppose that both F and G are purely discontinuous, i.e. that f and g are elements of H_d^2. In view of (11.8.1) we have

$$E(fg) = \Sigma_\alpha E(f_\alpha g_\alpha) = \Sigma_\alpha E(\Delta F_{T_\alpha} \Delta G_{T_\alpha}) = E(\Sigma_s \Delta F_s \Delta G_s)$$

in the notation of Corollary 1. (Notice that in a summable family of orthogonal elements of $L^2(P)$ only countably many elements can be nonzero).

Consequently for every stopping time S

$$E(\Sigma_{s \leq S} \Delta F_s \Delta G_s) = E(F_S G_S)$$

holds, which proves that $F_t G_t - W_t$ is a martingale in the case that f and g are elements of H_d^2. In the general case we get the result by writing F as sum of a continuous martingale and a purely discontinuous martingale.

Using Hölder's inequality and Schwartz's inequality, the fact that $W = (W_t)$ is of integrable variation follows from

$$E(\Sigma_s |\Delta F_s \Delta G_s|) = E(\Sigma_\alpha |\Delta F_{T_\alpha} \Delta G_{T_\alpha}|) \leq \Sigma_\alpha \sqrt{E[(\Delta F_{T_\alpha})^2]} \sqrt{E[(\Delta G_{T_\alpha})^2]}$$

$$\leq 4 \Sigma_\alpha \|f_\alpha\|_2 \|g_\alpha\|_2 \leq 4 \sqrt{\Sigma_\alpha (\|f_\alpha\|_2)^2} \sqrt{\Sigma_\alpha (\|g_\alpha\|_2)^2}$$

$$\leq 4 \|f\|_2 \|g\|_2 .$$

<u>Corollary 3</u>: The family of all $f \in L^2(P)$ such that the right continuous modification $F = (F_t)$ of $t \to E(f|F_t)$ is a process of integrable variation forms a dense subspace of H_d^2.

Proof: In view of 8.10 Corollary 3 every martingale of bounded variation is purely discontinuous, hence Corollary 3 follows from 11.8 Corollary 1.

§12 LOCALLY SUMMABLE AND SUMMABLE STOCHASTIC PROCESSES

In this section summability is studied and a simple criterion for a stochastic process to be summable is given. In view of the previous section, it is clear that every square integrable martingale $M = (M_t)$ is summable. Note however that the space $L^1(I_M)$ of I_M-integrable processes depends on whether M is considered as 2-summable process or as summable process, i.e. whether I_M is considered as measure with values in $L^2(P)$ or in $L^1(P)$.

The definition of local summability extends the notion of summability in such a way that every right continuous adapted stochastic process of bounded variation is locally summable.

From §2 we know that every stochastic process of integrable variation is summable, and consequently every quasimartingale, which is a sum of a square integrable martingale and an adapted process of integrable variation is summable. In order to introduce the concept of local summability, it is sufficient to know this result: this is the reason why we study local summability before summability.

The stochastic base in this section is fixed as in §11.

Let us first give a sufficient condition for a quasimartingale to be decomposed into a sum of a square integrable martingale and a process of integrable variation:

<u>12.1 Proposition</u>: Let $X = (X_t)$ be a quasimartingale, bounded in modulus by a constant K. Then X is the sum $X = M + V$ of a square integrable martingale M and a predictable process of integrable variation V.

<u>Proof</u>: According to Theorem 9.11 (Riesz decomposition theorem) X can be written as a sum $X = F + P$, where $F = (F_t)$ is a martingale and $P = (P_t)$

is a quasipotential. From the uniqueness of the Riesz decomposition we deduce that $F = (F_t)$ is the right continuous modification of $t \to E(f|F_t)$, with $f = \lim_{t \to \infty} X_t$, hence the martingale (F_t) is bounded by K and we therefore may assume that $X = (X_t)$ is a quasipotential, bounded in modulus by a constant K, and hence of class (D).

According to Theorem 9.15 and 9.15 Corollary X can be written as a sum $X = M + V$, where $V = (V_t)$ is a predictable process of integrable variation and the martingale $M = (M_t)$ is a right continuous modification of $t \to E(V_\infty|F_t)$. We show that V_∞ is an element of $L^2(P)$, which completes the proof.

Integration by parts yields $V_\infty^2 = \int (V_t + V_{t-})dV_t$, hence $V_\infty^2 = \int [(V_\infty - V_t) + (V_\infty - V_{t-})]dV_t$ holds.

We have $X_T = E(V_\infty - V_T|F_T)$ for every stopping time and $X_{T-} = E(V_\infty - V_{T-}|F_{T-})$ by virtue of Theorem 6.3 for every predictable stopping time T, therefore (Theorem 7.15) (X_t) is the well measurable projection of the process $(V_\infty - V_t)$ and (X_{t-}) is the predictable projection of $(V_\infty - V_{t-})$.

According to Theorem 8.8 $V = (V_t)$ commutes with the predictable projection and hence with the well measurable projection, which gives finally:

$$E(V_\infty^2) = E[\int (X_t - X_{t-})dV_t] \le 2 K E(|V|_\infty).$$

Corollary: Every quasimartingale, bounded in modulus by a constant K, is summable.

12.2 Definition: A right continuous adapted stochastic process $X = (X_t)$ is called *locally summable* if there is an increasing sequence (T_n) of stopping times T_n with $\lim_n T_n = \infty$, such that the process $\chi_{[0,T_n[} X$ is summable for every n.

A stopping time T is said to be *reducing* $X = (X_t)$ if the stopped process X^T is summable, or equivalently if the process $X_{[0,T]}X$ is summable.

A process $X = (X_t)$ is called a *local quasimartingale* if there is an increasing sequence (T_n) of stopping times T_n with $\lim_n T_n = \infty$, such that the stopped process X^{T_n} is a quasimartingale for every n.

A process $X = (X_t)$ is called a *semimartingale* if X can be written as a sum of local martingale and an adapted process of bounded variation.

Notice that all processes, defined above, are right continuous and have left hand limits.

<u>12.3 Theorem:</u> Let $X = (X_t)$ be a right continuous adapted stochastic process.

Consider the following assertions:

(a) X is locally summable.

(b) X is a semimartingale.

(c) X is a local quasimartingale.

(d)   There is an increasing sequence (T_n) of stopping times T_n reducing X with $\lim_n T_n = \infty$.

(e) X can be written as a sum $X = M + V$ of a local martingale $M = (M_t)$ and a predictable process of bounded variation $V = (V_t)$ zero at time $t = 0$.

The following implications hold: (a) $\Leftrightarrow$ (b) and (c) $\Leftrightarrow$ (d) $\Leftrightarrow$ (e) $\Rightarrow$ (b). If X is predictable we have moreover (b) $\Rightarrow$ (c). Furthermore the decomposition in (e) is unique.

<u>Proof:</u> (e) $\Rightarrow$ (b) is obvious.

(b) $\Rightarrow$ (a): Let $X = M + V$ be the sum of the local martingale M and the process of bounded variation V.

Define the increasing sequence (S_n) of stopping times by

$$S_n(\omega) = \inf\{t \in R_+ : |M_t(\omega)| \geq n \text{ or } |V|_t(\omega) \geq n\}.$$

Since $M = (M_t)$ and $|V| = (|V|_t)$ are right continuous processes with left hand limits, we have $\lim_n S_n = \infty$.

Furthermore the stopped process M^{S_n} is a local martingale of class (D) and therefore a uniformly integrable martingale by virtue of 9.15 Corollary for every n. Consequently

$$\chi_{[0,S_n[}M = M^{S_n} - M_{S_n}\chi_{[S_n,\infty[}$$

is as a difference of a uniformly integrable martingale and an adapted process of integrable variation a quasimartingale, which is moreover bounded in modulus by n and hence summable according to 12.1 Corollary. On the other hand $\chi_{[0,S_n[}V$ is clearly an adapted process of integrable variation and therefore $\chi_{[0,S_n[}X = \chi_{[0,S_n[}(M+V)$ is summable for every n, which proves that X is locally summable.

(c) $\Rightarrow$ (d): Let (T_n) be an increasing sequence of stopping times with $\lim_n T_n = \infty$, such that the stopped process X^{T_n} is a quasimartingale for every n. Define the sequence (U_n) of stopping times U_n by

$$U_n(\omega) = \{t \in R_+ : |X_t(\omega)| \geq n\},$$

and the sequence (S_n) by $S_n = U_n \wedge T_n$. We have $\lim_n S_n = \infty$ and $X^n = X^{S_n}$ is a quasimartingale for every n. Hence $X^n_{S_n}$ is an element of $L^1(P)$ and, since X^n is bounded in modulus by n on $[0, S_n[$,

$$X^n = \chi_{[0,S_n[}X^n + X^n_{S_n}\chi_{[S_n,\infty[}$$

is the sum of a bounded quasimartingale and an adapted process of integrable variation. By virtue of 12.1 Corollary X^n is summable for every n, which proves the implication (c) $\Rightarrow$ (d).

127

(d) ⇒ (e): for every n the summable process $X^n = X^{T_n}$ is a quasi-martingale of class (D) according to Theorem 11.2 and has therefore a unique decomposition into a uniformly integrable martingale M^n and a predictable process V^n of integrable variation, zero at time $t = 0$, by virtue of Theorem 9.15 and 9.15 Corollary. Consequently for $m \geq n$ the processes M^m and V^m, stopped at T_n, yield M^n and V^n. The (pointwise) limits

$$M = \lim_n M_n \quad \text{and} \quad \lim_n V^n = V$$

therefore exist on the complement of an evanescent subset of $R_+ \times \Omega$, and $X = M + V$ is the desired decomposition of X. (e) ⇒ (c): every local martingale is a local quasimartingale, it therefore suffices to show that every predictable process $V = (V_t)$ of bounded variation is a local quasi-martingale. Since V is predictable, the stopping time T_n, defined by $T_n(\omega) = \inf \{t \in R_+ : |V|_t(\omega) \geq n\}$, is predictable by 7.6 Corollary 3 for every n, and $P(\{T_1 > 0\}) = 1$ because of $V_0 = 0$. Let for fixed n $(S_{n,m})$ be an increasing sequence of stopping times $S_{n,m}$ satisfying $P(\{S_{n,m} < T_n\}) = 1$ for every m and $\lim_m S_{n,m} = T_n$. Define U_k as $U_k = \sup \{S_{n,m} : n \leq k \text{ and } m \leq k\}$. We have $\lim_k U_k = \infty$ and the stopped process $|V|^{U_k}$ is bounded in modulus by k for every k. This proves that V is a local quasimartingale.

a) ⇒ b): Define the sequence (T_n) of stopping times T_n by induction:

$$T_1(\omega) = \inf \{t \in R_+ : |X_t(\omega) - X_{t-}(\omega)| \geq 1\}$$

and for $n \geq 2$

$$T_n(\omega) = \inf \{t > T_{n-1}(\omega) : X_t(\omega) - X_{t-}(\omega)| \geq 1\},$$

where $T_n(\omega) = \infty$ if this set is empty or if $T_{n-1}(\omega) = \infty$. Since X is right continuous, we have $T_n > T_{n+1}$ on the set $A_n = \{T_n < \infty\}$, moreover a right continuous path with left hand limits can only have a finite number of jumps of size greater than 1, on every closed bounded interval therefore the process $W = \chi_{A_n} \Delta X_{T_n} \chi_{[T_n, \infty[}$ is a process of bounded variation which is locally summable in view of the implication b) ⇒ a), which is already proved.

Consequently $Y = X - W$ is locally summable and satisfies $|\Delta Y_t(\omega)| \leq 1$ for all (t, ω). As in the proof of the implication c) ⇒ d) it can be shown that Y satisfies d), and because d) implies e), Y can be written as sum $Y = M + V$ of a local martingale M and a predictable process V of bounded variation. Hence $X = M + V + W$ is a semimartingale, which proves the implication a) ⇒ b).

Using 7.6 Corollary 3, it is easily proved by induction, that the stopping times T_n, defined in the proof of the implication a) ⇒ b) are predictable in the case where X is predictable, and hence $\chi_{A_n} \Delta X_{T_n}$ is F_{T_-}-measurable according to Theorem 7.8. Consequently W is a predictable process in this case which proves b) ⇒ e) and hence b) ⇒ c) in the case where X is predictable.

That the decomposition in e) is unique follows from Theorem 9.12.

Corollary: If $X = (X_t)$ is a local quasimartingale, f and element of $L^\infty(P)$ and $F = (F_t)$ the right continuous modification of $t \to E(f|F_t)$. There is a unique predictable process $<X,F> = (<X,F>_t)$ of bounded variation, such that $(X_t F_t - <X,F>_t)$ is a local martingale, zero at time $t = 0$.

Proof: According to 12.3 c) there is an increasing sequence (T_n) of stopping times, such that the stopped process X^{T_n} is summable for every n. By Theorem 11.2 the process $(X_t F_t)$, stopped at T_n, is therefore a quasimartingale for every n. Hence $(X_t F_t)$ is a local quasimartingale, and the assertion follows from 12.3 e).

__12.4 Definition:__ Let $X = (X_t)$ be a summable stochastic process, $Z = (Z_t)$ a predictable process. Instead of saying that Z is integrable with respect to the measure I_X, generated by X, we shall shortly say in the sequel that Z *is integrable with respect to* X.

If X is a locally summable process, we shall call a predictable process Z *integrable with respect to* X, if the process $\chi_{\{Z>0\}} X$ is summable and Z is integrable with respect to $\chi_{\{Z>0\}} X$.

A predictable process Z is called *locally integrable with* respect to X if there is an increasing sequence (T_n) of stopping times T_n, satisfying

a) $\lim_n T_n = \infty$ and

b) Z is integrable with respect to the stochastic process $\chi_{[0,T_n[} X$ for every n.

Integrability of Z implies local integrability, and predictable processes Z which are locally bounded, i.e. which admit an increasing sequence (T_n) of stopping times T_n with $\lim_n T_n = \infty$, such that the process $\chi_{[0,T_n[} Z$ is bounded in modulus (by a constant) for every n, are locally integrable with respect to every locally summable process X.

The integral $I_X(Z)$ of a predictable process Z, which is integrable with respect to a locally summable process X, is a well defined element of $L^1(P)$.

If Z is locally integrable but not integrable, the integral has no sense as an integral with respect to a $L^1(P)$-valued measure. However, in the particular case, that $X = (X_t)$ is an adapted process of bounded variation, the integral $\int_A \chi_A Z_t \, dX_t$, defined in §2, is a well defined F-measurable real function on Ω, at least for locally bounded processes

Z and predictable sets A contained in a stochastic interval $[0,T]$, where T is a finite stopping time.

To be able to define the integral in the general case, we need the following

12.5 Proposition: Let $X = (X_t)$ and $Y = (Y_t)$ be two summable processes and $Z = (Z_t)$ be integrable with respect to X and Y. If X and Y coincide on a set $F \in \mathcal{F}$, then $I_X(Z)$ and $I_Y(Z)$ are (P-almost surely) equal on F.

Proof: Let G be an $\mathcal{F}$-measurable subset of F. We have

$$E(\chi_G I_X(Z)) = \int Z \, d I'_X(\chi_G) \text{ and } E(\chi_G I_Y(Z)) = \int Z \, d I'_Y(\chi_G).$$

The proof of the proposition follows now from the following

Lemma: Let f be an element of $L^\infty(P)$ and suppose that $fX = fY$ holds. Then $I'_X(f)$ and $I'_Y(f)$ coincide.

Proof of the Lemma: We have

$$I'_X(f)([0_H]) = E(\chi_H X_0 f) = E(\chi_H Y_0 f) = I'_Y(f)([0_H])$$

for every H of $\mathcal{F}_0$, and

$$I'_X(f)(]S,T]) = E[(X_T - X_S)f] = E[(Y_T - Y_S)f] = I'_Y(f)(]S,T]),$$

for every pair of stopping times $S \leq T$. Since stochastic intervals of this form are a generator of the predictable σ-algebra Σ_p, which is stable under finite intersections, this implies $I'_X(f) = I'_Y(f)$.

Corollary: Let X be a locally summable process and let Z be a predictable process, integrable with respect to $\chi_{[0,T[}X$ and $\chi_{[0,S[}X$ for two stopping times T and S. Then $I_{\chi_{[0,T[}X}(Z)$ and $I_{\chi_{[0,S[}X}(Z)$ coincide on the set

$\{T \leq S\}$.

<u>12.6 Definition</u>: A predictable set $A \in \Sigma_p$ is called *finite* if A is contained in a stochastic interval $[0,T]$, where T is a finite stopping time.

Let now X be a locally summable process and let Z be locally integrable with respect to X. For a finite set $A \in \Sigma_p$ we define the integral of $\chi_A Z$ with respect to X as the limit (in P-measure) of the sequence $(I_{Y^n}(\chi_A Z))$, where Y^n is the process $Y^n = \chi_{[0,T_n[} X$ $(n=1,2,\ldots)$ and (T_n) is a sequence of stopping times, satisfying 12.4 a) and b).

In view of 12.5 Corollary, the limit above exists P-almost surely and hence in probability, and is independent of the particular choice of the sequence (T_n).

For a finite set $A \in \Sigma_p$ we write again $I_X(\chi_A Z)$ for the integral of $\chi_A Z$ with respect to X, which is a (up to a P-null set) well defined real function, provided X is locally summable and Z is locally X-integrable. In particular the stochastic interval $[0,t]$ is a finite set for every $t \in R_+$ and $t \to I_X(\chi_{[0,T]} Z)$ defines an $M(F)$-process, where $M(F)$ denotes the vector space of real F-measurable functions on Ω (see §1). Taking into account 12.5 Corollary and Theorem 11.4, it is easily seen that $t \to I_X(\chi_{[0,t]} Z)$ has a right continuous modification $X(Z) = (X(Z)_t)$, which will be called *the stochastic integral of Z with respect to X*, as in 11.5.

Let us list some of the properties of that integral in the following

<u>12.7 Theorem</u>: Let $X = (X_t)$ be a locally summable process and $Z = (Z_t)$ locally integrable with respect to X.

a) For every predictable finite subset A of $R_+ \times \Omega$ the integral $I_X(\chi_A Z)$ of $\chi_A Z$ with respect to X is an up to a P-null set defined real function on Ω, which is measurable with respect to $F_{D_{A^c}}$, where D_{A^c} is the debut of the set A^c.

b) The $M(F)$-process $t \to I_X(\chi_{[0,t]} Z)$ has a right continuous modification $X(Z) = (X(Z)_t)$, which is a semimartingale.

c) If Z is integrable, $(X(Z)_t)$ is a quasimartingale of class (D), if in addition X is a local martingale, $(X(Z)_t)$ is a uniformly integrable martingale.

d) Provided there is an increasing sequence (T_n) of stopping times with $\lim_n T_n = \infty$, such that $\chi_{[0,T_n]} Z$ is integrable with respect to X for every n, then $(X(Z)_t)$ is a local quasimartingale.

If moreover X is a local quasimartingale, then $X(Z) = (X(Z)_t)$ is the unique local quasimartingale, satisfying

$$<X(Z),F>_t = \int \chi_{[0,t]} Z_s d<X,F>_s \quad \text{for every } f \in L(P) \;.$$

Remark: The definition of $<X,F> = (<X,F>_t)$ in the case that X is a local quasimartingale and f an element of $L^\infty(P)$ is given in 12.3 Corollary.

Proof: After the argument preceding the theorem, the proof of the remaining assertions is easily accomplished by using theorems 11.4, 12.3 and 12.3 Corollary.

In the remainder of this section we shall study stochastic processes which are not only locally summable but summable. In Theorem 11.2 the vector space of summable processes was characterized as a subspace of the vectorspace of quasimartingales, we shall define a norm on the space of summable processes and give a condition for a quasimartingale to be summable.

Let us denote by SP the vector space of summable processes. As usual we identify indistinguishable processes. By SM we denote the subspace of SP, formed by summable martingales. Recall that pSV is the space of all predictable processes of integrable variation, and that pSV_0 is the subspace of pSV, consisting of all $V = (V_t) \in pSV$ such that $V_0 = 0$ holds. Clearly pSV is a subspace of SP.

SP is a normed space under the norm $\|X\| = svI_X(R_+ \times \Omega)$, where svI_X is the semivariation of the $L^1(P)$-valued measure I_X generated by $X \in SP$.

12.8 Theorem: Let SP be the normed vector space of summable processes. The following assertions hold:

a) SP is a Banach space under the norm $\|X\| = svI_S(R_+ \times \Omega)$.

b) Every element of SP is a quasimartingale of class (D) and SP contains all square integrable martingales and all quasi-martingales, bounded in modulus by a constant K.

c) Denote by $X = M_X + V_X$ the unique decomposition of X into an element $M_X \in SM$ and $V_X \in pSV_0$, given by Theorem 9.12. Then $X \to M_X$ and $X \to V_X$ are continuous linear projections on SM, $X \to V_X$ is of norm ≤ 1 and consequently $X \to M_X$ is of norm ≤ 2.

d) SM and pSV_0 are closed linear subspaces of SP.

Before proving the theorem, we prove the following lemma, which is itself of interest in the context of §2.

12.9 Lemma: Let $V = (V_t) \in pSV_0$ be a predictable process of integrable variation. The semivariation svI_V and variation vI_V of the $L^1(P)$-valued measure I_V, generated by V coincide and are equal to the variation $|\mu_V|$ of the real measure μ_V, generated by V on Σ_p.

Proof: In order to prove the lemma it suffices to prove that $sv I_V$ is equal to $|\mu_V|$, because that implies that the semivariation is additive and therefore coincides with the variation of I_V.

Let $A \in \Sigma_p$ be a predictable set. We have

$$|\mu_V|(A) = \sup\{\int Z d\mu_V : Z \in B(\Sigma_p), |Z| \leq \chi_A\}$$

$$= \sup\{E(I_V(Z)) : Z \in B(\Sigma_p), |Z| \leq \chi_A\}$$

$$\leq sv I_V(A)$$

and $sv I_V(A) = \sup\{\|I_V(Z)\|_1 : Z \in B(\Sigma_p), |Z| \leq \chi_A\}$

$$\leq \sup\{E(I_{|V|}(Z)) : Z \in B(\Sigma_p), |Z| \leq \chi_A\}$$

$$= E(\int \chi_A d|V|_t) = |\mu_V|(A)$$

according to Theorem 2.6.

Proof of the theorem: That assertion b) holds is a consequence of 12.1 Corollary.

We next prove c). For every pair of stopping times $S \leq T$

$E[I_X(]S,T])] = E[X_T - X_S] = E[V_T - V_S] = \mu_V(]S,T])$ holds according to 9.13 Corollary 2 and consequently we have $E[I_X(\chi_{]0,\infty[}Z)] = \int Z d\mu_V$ for every predictable bounded process $Z \in B(\Sigma_p)$, provided $X = M + V$ is the decomposition of the element $X \in SP$ into a uniformly integrable martingale M and a predictable process of integrable variation $V \in pSV_0$. There is a process $Z \in B(\Sigma_p)$, satisfying $|Z| \leq 1$ and $\chi_{[0]} Z = 0$, such that $|\mu_V|(R_+ \times \Omega) = \int Z d\mu_V$ and we therefore get

$|\mu_V|(R_+ \times \Omega) = \int Z d\mu_V = E[I_X(Z)] \leq sv I_X(R_+ \times \Omega)$ and the equality

$svI_V(R_+ \times \Omega) = |\mu_V|(R_+ \times \Omega)$, given by Lemma 12.9, proves assertion c). Assertion d) is an immediate consequence of c).

In order to prove a), we show that the subspaces pSV_0 and SM of SP are complete.

In the case of pSV_0 this follows immediately from Lemma 12.9, because we know that pSV_0 is a Banach space under the norm $V \to |\mu_V|(R_+ \times \Omega)$, (Theorem 8.1).

In the case of SM, we notice that every element $M = (M_t)$ of SM is a uniformly integrable martingale, hence (M_t) tends to an element M_∞ of $L^1(P)$ as t tends to ∞, and $M = (M_t)$ is the right continuous modification of $t \to E(M_\infty | F_t)$ (see Theorem 5.4). We therefore can embed SM into $L^1(P)$ and clearly this embedding is continuous. Assertion a) is now a consequence of the following proposition:

12.10 Proposition: Consider SM as subspaces of $L^1(P)$ via the embedding map which maps $F = (F_t) \in SM$ to the element $f = F_\infty = \lim_{t \to \infty} F_t$. The unit ball SM_1 of SM is contained in the unit ball of $L^1(P)$ and SM_1 is closed in $L^1(P)$.

Corollary: SM is a Banach space.

Proof: It is clear that the unit ball SM_1 of SM is contained in the unit ball of $L^1(P)$.

Let (f_n) be a sequence of elements of $L^1(P)$, which converges (in the $L^1(P)$-norm) to an element $f \in L^1(P)$ and satisfies $f_n \in SM_1$ for every n. We have to show that the right continuous modification $F = (F_t)$ of $t \to E(f | F_t)$ is an element of the unit ball of SP. Let an $\varepsilon > 0$, an element H of F_0, a finite sequence $a_0, a_1, \ldots, a_k$ of real numbers $|a_i| \leq 1$ and an increasing sequence $T_1, T_2, \ldots, T_{k+1}$ of stopping times be given. Denote by $F^n = (F^n_t)$ the right continuous modification of $t \to E(f_n|F_t)$ ($n=1,2,\ldots$). There is an integer m such that $\|F^m_{T_i} - F_{T_i}\|_1 \leq \varepsilon/k$ and $\|F^m_o - F_o\|_1 \leq \varepsilon$ holds for all $i=1,\ldots,k+1$.

Consequently we have

$$\|\Sigma^k_{i=1} a_i(F_{T_{i+1}} - F_{T_i}) + a_o \chi_H F_o\|_1$$

$$\leq 3\varepsilon + \|\Sigma^k_{i=1} a_i(F^m_{T_{i+1}} - F^m_{T_i}) + a_o \chi_H F^m_o\|_1 \leq 3\varepsilon + 1.$$

This proves that the measure I_F generated by $F = (F_t)$ on $\mathcal{A}$ (see 11.1) has a semivariation $svI_F(R_+ \times \Omega) \leq 1$. That $F = (F_t)$ is an element of SP with norm ≤ 1 is now a consequence of the following Lemma, because $F = (F_t)$ is a uniformly integrable martingale and hence of class (D).

The following lemma is also needed for the proof of Proposition 12.12.

12.11 Lemma: Let $X = (X_t)$ be a right continuous adapted stochastic process. Consider the following assertions:

i) The process X is of class (D) and the stochastic measure I_X, generated by X by $\mathcal{A}$ (see 11.1) is bounded.

ii) X is of class (D) and for every $f \in L^\infty(P)$ the process $XF = (X_t F_t)$, where $F = (F_t)$ is the right continuous modification of $t \to E(f|F_t)$, is a quasimartingale.

iii) X is summable.

The following implications hold: i) $\Rightarrow$ ii) $\Rightarrow$ iii).

Proof: i) $\Rightarrow$ ii): I_X is bounded if and only if I_X is of bounded semivariation, hence the implication i) $\Rightarrow$ ii) is an immediate consequence of the definition of the semivariation and Theorem 9.2.

ii) $\Rightarrow$ iii): According to Theorem 11.2 the process X is summable if and only if $XF = (X_t F_t)$ is a quasimartingale of class (D) for every $f \in L^\infty(P)$. Now the set $\{X_T : T$ is a finite stopping time$\}$ is uniformly integrable and, since $|F_T|$ is bounded by a constant K uniformly for every stopping time T, the set $\{X_T F_T : T$ is a finite stopping time$\}$ is uniformly integrable as well. This proves that XF if of class (D) for every $f \in L^\infty(P)$.

The following proposition characterizes elements of SM:

12.12 Proposition: Let $M = (M_t)$ be a right continuous martingale and denote by M^* the random variable $M^* = \sup_{t \in R_+} |M_t|$. M is an element of SM if and only if M^* is integrable.

<u>Proof</u>: We first prove that the condition is sufficient. Let $T_1, T_2, \ldots, T_n$ be an increasing sequence of stopping times and $a_1, a_2, \ldots, a_n$ a finite sequence of real numbers, satisfying $|a_j| \leq 1$ for $j = 1, 2, \ldots, n$. By virtue of 4.18 Corollary 2 we have $\|\Sigma_{j=2}^n a_j (M_{T_j} - M_{T_{j-1}}) + a_1 M_{T_1}\|_1 \leq 16 \, E(M^*)$.

By virtue of Lemma 12.11 this inequality proves that $M = (M_t)$ is summable and hence an element of SM. In order to prove the converse, let us first recall that the sequence (r_k) of Rademacher functions is a sequence of real functions, defined on the interval $[0,1]$ as $r_k(s) = \text{sign} \sin 2^k \pi s$ ($k=1,2,\ldots$). A particular case of the left hand side of Khintchin's inequality yields: There is a constant K such that for any finite sequence $\beta_1, \beta_2, \ldots, \beta_n$ of real numbers the following inequality holds:

$$K(\Sigma_{k=1}^n |\beta_k|^2) \leq \int_0^1 |\Sigma_{k=1}^n \beta_k r_k(s)| ds.$$

The constant can be taken as $K = 1/8$, and we shall use this value for K in sequel (see [14]).

Suppose now that $M = (M_t)$ is an element of SM with norm ≤ 1, and let an increasing sequence $T_1, T_2, \ldots, T_n$ of stopping times be given. Define the sequence $f_1, f_2, \ldots, f_n$ of elements $f_k \in L^1(P)$ by $f_1 = M_{T_1}$ and

$$f_k = M_{T_k} - M_{T_{k-1}} \quad \text{for} \quad k=2,3,\ldots,n.$$

Since $M = (M_t)$ is an element of SM of norm ≤ 1, we have

$$E[|\Sigma_{k=1}^n f_k r_k(s)|] \leq svI_M(R_+ \times \Omega) \leq 1 \quad \text{for every} \quad s \in [0,1] \quad \text{and} \quad n = 1, 2, \ldots .$$

According to the Khintchin inequality this yields by virtue of Fubini's Theorem:

$$E[(\Sigma_{k=1}^n |f_k|^2)^{1/2}] \leq 8 \, E[\int_0^1 |\Sigma_{k=1}^n f_k r_k(s)| ds]$$
$$= 8 \int_0^1 E[|\Sigma_{k=1}^n f_k r_k(s)|] ds$$
$$\leq 8 .$$

Since the sequence $T_1, T_2, \ldots, T_n$ was arbitrary, we get $E(M^*) \leq 40$ by virtue of 4.18 Corollary 1. This completes the proof.

The vector space of all $f \in L^1(P)$ such that the right continuous modification $F = (F_t)$ of $t \to E(f|F_t)$ satisfies $E(\sup_{t \in R_+} |F_t|) < \infty$

is usually denoted by H^1 (see §4). H^1 is a Banach space under the norm $f \to E(\sup_{t \in R_+} |F_t|)$. According to Theorem 4.11 $L^p(P)$ is a subspace of H^1, and the embedding of $L^p(P)$ into H^1 is continuous. On the other hand the unit ball of H^1 is obviously contained in the unit ball of $L^1(P)$, so that the embedding of H^1 into $L^1(P)$ is a contraction. Examining the proof of the previous proposition, we get the following Corollary:

<u>Corollary</u>: The Banach spaces H^1 and SM are isomorphic under the linear operator which maps an element $f \in H^1$ to the right continuous modification $F = (F_t)$ of $t \to E(f|F_t)$. If we denote the norm on H^1 by p_1, we get

$p_1(f) \leq 40 \|F\|$, and $\|F\| \leq 16 \, p_1(f)$.

In the following theorem several equivalent conditions for a quasi-martingale to be summable are listed:

<u>12.13 Theorem</u>: Let $X = (X_t)$ be an adapted right continuous stochastic process. If we denote by I_X the stochastic measure, generated by X on A (see 11.1), the following assertions are equivalent:

a) X is summable

b) the additive measure I_X is (weakly) σ-additive and bounded on A.

c) $X = (X_t)$ is a quasimartingale, satisfying $E(\sup_{t \in R_+} |X_t|) < \infty$.

d) the process $X = (X_t)$ is of class (D) and the measure I_X is bounded on A.

e) the process $X = (X_t)$ is of class (D) and for every $f \in L^\infty(P)$ the process $XF = (X_t F_t)$, where $F = (F_t)$ is a right continuous modification of $t \to E(f|F_t)$, is a quasimartingale.

<u>Proof</u>: In view of Theorem 10.10 and the weak sequential completeness of $L^1(P)$, the equivalence of a) and b) is obvious.

a) $\Rightarrow$ c): According to 12.8 c) X can be written as a sum $X = M + V$, where the martingale $M = (M_t)$ is an element of SM and $V = (V_t)$ is a process of integrable variation. M satisfies c) according to Proposition 12.12 and since it is clear that V satisfies c) the quasimartingale X satisfies c) too.

c) $\Rightarrow$ d) Clearly c) implies that X is of class (D), the assertion d) is therefore an easy consequence of the Doob-Meyer decomposition theorem for quasimartingales (9.15) and Proposition 12.12.

The implications d) ⇒ e) ⇒ a) finally are proved in Lemma 12.11.

Remark: Notice however that, in contrast to 12.12 Corollary, the norm $X \to E(\sup_t |X_t|)$ is not equivalent to the SP-norm on SP, as the following example shows: If Ω consists of one point, SP is the space of right continuous functions of bounded variation on R_+ (the martingales are the constant functions) and the SP-norm is the norm of the total variation. This norm is obviously not equivalent to the sup-norm.

§13 Decomposition of SM and representation of martingales as stochastic integrals

In §12 the Banach space of summable martingales was introduced and according to Proposition 12.10 and 12.12 Corollary we can identify SM with the Banach space H^1 of all $f \in L^1(P)$ such that the right continuous modification $F = (F_t)$ of $t \to E(f|F_t)$ satisfies $E(\sup_t |F_t|) < \infty$.

In this section we study the space SM and the results will be applied to questions related to the representation of martingales as stochastic integrals.

Let us first note the following theorem, the proof of which is left to the reader (compare with Theorem 11.6):

13.1 Theorem: Let the martingale $M = (M_t)$ be an element of SM. For every M-integrable process $Z \in L^1(I_M)$, $I_M(Z)$ is an element of H^1, and if we equip H^1 with the (equivalent) SM-norm, the integral operator $Z \to I_M(Z)$ is an isometry from $L^1(I_M)$ into H^1.

Corollary 1: The image of $L^1(I_M)$ under I_M is a closed subspace of H^1.

Corollary 2: The set function $A \to I_M(A) = I_M(\chi_A)$ on Σ_p is a σ-additive measure with values in H^1. In particular the adjoint I_M' of the operator $I_M : B(\Sigma_p) \to H^1$ maps the dual of H^1 into $ca(\Sigma_p)$, or more precisely, into $sca(\Sigma_p)$.

The following proposition shows that the martingale convergence theorem holds in SM:

13.2 Proposition: Let $F = (F_t)$ be a summable martingale and (T_n) an increasing sequence of stopping times. The sequence (F^n) of the stopped martingales $F^n = (F_{t \wedge T_n})$ converges in SM to a martingale $G = (G_t) \in $ SM, and we have $F^n = G_{t \wedge T_n}$ for every $t \in R_+$ and $n = 1, 2, \ldots$

Proof: The SM-norm of the martingale $F^n - F^m$ ($n \geq m$) is equal to the semivariation $svI_F([T_m, T_n])$, where I_F is the stochastic measure, generated by F on Σ_p. Since I_F is σ-additive, the claimed convergence follows from 12.10 Corollary. The last assertion of the proposition is a consequence of the Optional Stopping Theorem for martingales.

Remark: In terms of H^1 (equipped with the SM-norm), the proposition above reads:

For every stopping time T the conditional expectation $f \to E(f|F_T)$ is a projection of norm ≤ 1 on H^1, and for every increasing sequence of stopping times T_n, the sequence $(E(f|F_{T_n}))$ converges to an element $g \in H^1$, satisfying $E(g|F_T) = E(f|F_{T_n \wedge T})$ for every stopping time T.

Corollary: $L^\infty(P)$ and $L^2(P)$ are dense subspaces of H^1.

Proof: In view of the argument previous to 12.12 Corollary it suffices to show that $L^2(P)$ is dense. Let f be an element of H^1 and let an $\varepsilon > 0$ be given. Define the sequence (T_n) of stopping times T_n by
$$T_n(\omega) = \inf\{t > 0 : |F_t|(\omega) \geq n\},$$
where $F = (F_t)$ is a right continuous modification of $t \to E(f|F_t)$. We have $\sup_n T_n = \infty$ and consequently there is a m such that $g = E(f|F_{T_m})$ satisfies $\|f - g\| \leq \varepsilon$ in the SM-norm.

Write $G = (G_t)$ for the right continuous modification of $t \to E(g|F_t)$. According to Proposition 12.1 the quasi-martingale $X = X_{[0, T_m[} G$ has a decomposition $X = M + V$, where M is a square integrable martingale and $V \in pSV_0$ is a predictable process of integrable variation.

The process $Y = g\chi_{[T_m, \infty[} = L_{T_m} \chi_{[T_m, \infty[}$ is an element of wSV and there is a bounded F_{T_m}-measurable random variable h such that $(g-h)\chi_{[T_m, \infty[}$ has a

wSV-norm smaller than ε. Observing that the dual predictable projection of $gX_{[T_m,\infty[}$ is the process $-V$, we conclude that the wSV-norm of $(g-h)X_{[T_m,\infty[} + V + W$ is smaller than 2ε, if W denotes the dual predictable projection of $hX_{[T_m,\infty[}$.

According to Proposition 12.1 the martingale $N = \widehat{hX_{[T_m,\infty[}}^{-W}$ is square integrable, and, since the SM-norm is smaller than the wSV-norm, we have $\|gX_{[T_m,\infty[} + V - N\| \le 2\varepsilon$ in the SM-norm.

Thus we get finally

$$\|f - (M_\infty + N)\| = \|F - G + (G - M - N)\|$$
$$\le \|f - g\| + \|G - X_{[0,T_m[}G + V - N\|$$
$$\le \varepsilon + \|gX_{[T_m,\infty]} + V - N\| \le 3\varepsilon$$

in the SM-norm, which completes the proof of the Corollary.

<u>13.3 Definition</u>: Let $X = (X_t)$ be a locally summable stochastic process. A stochastic process $Y = (Y_t)$ is called representable as a stochastic integral with respect to X if there is a locally integrable (predictable) process $Z = (Z_t)$, such that Y is the stochastic integral $X(Y) = (X(Z)_t)$ of Z with respect to X (see Definition 12.6).

The question, which quasimartingales are representable as a stochastic integral with respect to X is easily settled if X is a predictable process of integrable variation. If X is a predictable process of integrable variation, a quasimartingale Y is representable as a stochastic integral with respect to X, if and only if Y is a predictable process of integrable variation such that the real measure μ_Y, generated by Y on Σ_p, is absolutely continuous with respect to μ_X. In view of §2 and §8 the assertion above is a consequence of the Radon Nikodym Theorem.

The following theorem lists conditions, under which for a given local martingale $X = (X_t)$ every local martingale $Y = (Y_t)$ is representable as a stochastic integral with respect to X.

13.4 Theorem: Let $X = (X_t)$ be a local martingale. The following assertions are equivalent:

a) Every local martingale is representable as a stochastic integral with respect to X.

b) Every summable martingale is representable as a stochastic integral with respect to X.

c) Every right continuous martingale, bounded in modulus by a constant, is representable as a stochastic integral with respect to X.

Proof: Given a local martingale $X = (X_t)$, there is an increasing sequence (T_n) of stopping times T_n with $T_0 = 0$ and $\lim_n T_n = \infty$, such that the stopped process $X^n = (X_{t \wedge T_n})$ is a summable martingale for every n. Put $\beta_n = (2^n \|X^{n+1} - X^n\|)^{-1}$, the norm being the SM-norm. The predictable process $Z = \Sigma_n \beta_n X^n \mathbf{1}_{]T_n, T_{n+1}]}$ is integrable with respect to X, the stochastic integral of Z with respect to X is a summable martingale $M = (M_t)$ (of SM-norm $\|M\| \leq 1$) and X is the stochastic integral of Z^{-1} with respect to M. We may therefore assume without loss of generality that X is a summable martingale. From the preceding argument we deduce moreover the equivalence of a) and b).

That b) implies c) is clear. On the other hand, b) means that the integral operator I_X maps $L^1(I_X)$ onto H^1. By virtue of 13.1 Corollary 1 this is equivalent to the assertion that the range of I_X is dense in H^1. Taking into account 13.2 Corollary, this completes the proof of the theorem.

As we saw in the proof of the previous theorem, denseness of the range of $B(\Sigma_p)$ in H^1 under the integral operator I_X is equivalent to the assertion, that every local martingale is representable as a stochastic integral with respect to the summable martingale X. Now a continuous linear operator has dense image if and only if the adjoint operator is injective, and we will show that it is in fact sufficient to know, that the adjoint I'_X of I_X, restricted to $L^\infty(P)$, is injective.

Some facts about the dual space of H^1, which will be proved later, are needed for this purpose:

The dual space of H^1 can be identified with a subspace BMO of $L^2(P)$, satisfying:

i) If f is an element of H^1 and g an element of BMO and if we denote by $F = (F_t)$ resp. $G = (G_t)$ the right continuous modifications of $f \to E(f|F_t)$ resp. $g \to E(g|F_t)$, then the process $FG = (F_t G_t)$ is a local quasimartingale.

ii) For every $g \in$ BMO there is a constant K, such that
$|\Delta G_T| = |G_T - G_{T_-}|$ is bounded by K uniformly for all stopping times T.

In particular for every $f \in H^1$ and $g \in$ BMO there is a unique predictable process $V = (V_t)$ of bounded variation, such that $FG - V$ is a local martingale, zero at time $t = 0$. We shall denote this process V by $<F,G> = (<F,G>_t)$ (see Theorem 12.3), and we have:

iii) $<F,G>$ is a process of integrable variation for every $f \in H^1$ and $g \in$ BMO, and the dual bilinear form on the pair (H^1, BMO) is given by $<f,g> = E(<F,G>_\infty)$.

It is clear that for an element $x \in SM$ the adjoint I'_X of the integral operator $I_X : B(\Sigma_p) \to H^1$ maps elements $g \in BMO$ to the real measure $\mu_{<X,G>} \in sca(\Sigma_p)$, generated by $<X,G>$ on Σ_p.

13.5 Theorem: Let $X = (X_t)$ be a summable martingale. The following assertions are equivalent:

a) Every local martingale is representable as a stochastic integral with respect to X.

b) The integral operator I_X maps the space $B(\Sigma_p)$ of bounded predictable processes onto a dense subspace of $L^1(P)$.

c) $<X,G> = 0$ for a right continuous martingale $G = (G_t)$, which is bounded in modulus by a constant K, implies $G = 0$.

d) $<X,G> = 0$ for an element $g \in BMO$ implies $g = 0$.

Proof: After the argument, preceding the theorem, it is sufficient to prove the implication c) $\Rightarrow$ d).

Suppose $<X,G> = 0$ holds for an element $g \in BMO$. By virtue of ii) above, the martingale $G = (G_t)$ is locally bounded, i.e. there is an increasing sequence (T_n) of stopping times with $\lim_n T_n = \infty$, such that the stopped martingale $G^n = (G_{t \wedge T_n})$ is bounded by a constant. Since for every n the process $<X,G^n>$ coincides with the process $<X,G>$ stopped at T_n, we have $G^n = 0$ for every n according to c). This implies $G = 0$ and hence $g = 0$, which completes the proof of c) $\Rightarrow$ d).

Corollary: Let $X = (X_t)$ be a summable martingale and suppose that F_0 is the σ-algebra, generated by all P-null sets. The following assertions are equivalent:

a) Every local martingale is representable as a stochastic integral with respect to X.

b) For every positive element $g \in L^\infty(P)$, such that $g\,dP$ is a probability measure, the following implication holds:
   If $X = (X_t)$ is a martingale with respect to $g\,dP$ then g is constant, equal to 1.

c) P is an extreme point of the set of all P-absolutely continuous probability measures P', such that $X = (X_t)$ is a martingale with respect to P'.

Remark: It is clear that the set of probability measures, mentioned in c) is convex. Since we are dealing with probability measures, we may omit the words 'P-absolutely continuous' in assertion c), or alternatively replace 'martingale' by 'local martingale'.

Proof: a) ⇒ b): If (X_t) is a $g\,dP$ martingale, then the process $(X_t G_t)$ is a martingale with respect to the probability measure P, where $G = (G_t)$ is the right continuous modification $t \to E(g|F_t)$ and the conditional expectations are taken with respect to P. Since F_0 is the trivial σ-field, G_0 is a constant function, and if we denote by $H = (H_t)$ the martingale $(H_t) = (G_t - G_0)$, we have $\langle X,H \rangle = 0$ and consequently $H = 0$ according to Theorem 13.5. Hence G_t is equal to the constant function G_0 for all t, which yields finally $g = G_0 = 1$ because $g\,dP$ is a probability measure.

b) ⇒ c): Let P' and P'' be probability measures, such that $X = (X_t)$ is a martingale with respect to P' and P'', and such that $P = \alpha P' + (1 - \alpha)P''$ holds for a real α satisfying $0 < \alpha < 1$. We have $P' \leq 1/\alpha\, P$ and hence $P' = g\,dP$ for an element $g \in L^\infty(P)$. According to b) this implies $g = 1$ and hence $P = P'$ i.e. P is an extreme point.

c) ⇒ a): Suppose a) does not hold. According to Theorem 13.5 there is a nonzero element g of $L^\infty(P)$, such that $<X,G> = 0$ holds or equivalently, such that $(X_t G_t)$ is a martingale, zero at time $t = 0$. We may suppose that the $L^\infty(P)$-norm of g is smaller than 1, i.e. that the measures $P' = (1+g)\,dP$ and $P'' = (1-g)\,dP$ are probability measures. We have $P' \neq P''$, (X_t) is a martingale with respect to P' and P'', and P is the convex combination $P = 1/2\,(P' + P'')$, which is absurd.

In the remainder of this section we shall prove a decomposition theorem for the case of summable martingales, similar to those of §11 for square integrable martingales. Furthermore the properties of BMO, used in the proof of Proposition 13.5, will be established.

13.6 Proposition: Let $M = (M_t)$ be a martingale. Put $\Delta M_t = M_t - M_{t-}$ for $t > 0$ and $\Delta M_0 = M_0$. There is a constant c, such that $E(\sqrt{\Sigma_s (\Delta M_s)^2}) \leq c\,\|M\|$ holds for every $M \in SM$, the norm being the SM-norm of M.

Proof: M_∞ is an element of H^1 (Proposition 12.12), suppose the H^1-norm $p_1(M_\infty)$ of M_∞ is smaller than 1, and let $t(n,0),\ldots,t(n,k_n)$ ($t(n,0) = 0$) be the n'th dyadic partition of the interval $[0,t]$ for a fixed $t \in R_+$. For every $S_n = [\Sigma_{i=0}^{k_n - 1} (M_{t(n,i+1)} - M_{t(n,i)})^2]^{1/2}$ satisfies $E(S_n) \leq 5$ according to 4.18 Corollary 1.

On the other hand $\Sigma_{s\leq t}(\Delta M_s)^2 \leq \liminf_n S_n^2$ holds, and the proposition follows from Fatou's lemma, Proposition 13.2 and 12.12 Corollary.

Corollary 1: The process $V = (V_t)$, defined by $V_t(\omega) = \Sigma_{s\leq t}(\Delta M_s(\omega))^2$ is a (well measurable) process of bounded variation, moreover $\sqrt{V} = (\sqrt{V_t})$ is of integrable variation.

We denote by dSM the closure in SM of the subspace $H_d^2 \subseteq SM$ of purely discontinuous square integrable martingales (c.f. 11.7). Elements of dSM are called purely discontinuous martingales.

Corollary 2: There are constants c_1 and c_2, such that
$$c_1 \|M\| \leq E(\sqrt{\Sigma_s (\Delta M_s)^2}) \leq c_2 \|M\|$$
holds for every purely discontinuous martingale $M \in dSM$.

Proof: The right hand side of the inequality is clear from the previous proposition.

Let $T_1, \ldots, T_n$ be a finite sequence of totally inaccessible or predictable stopping times, $f_1, \ldots, f_n$ a sequence of elements of $L^1(P)$, such that f_i is F_{T_i}-measurable for $i = 1, \ldots, n$.

Denote by V^i ($i=1,\ldots,n$) the compensating process of $W^i = f_i \cdot \chi_{[T_i, \infty[}$ and by M^i the martingale $M^i = W^i - V^i$ (c.f. 11.8). According to 11.8 Corollary 1 it suffices to establish the existence of the constant c_1 for martingales of the form $M = \Sigma_{i=1}^n M^i$, where $T_1, \ldots, T_n$ is a finite sequence of totally inaccessible or predictable stopping times with disjoint graphs. Recall that V^i is a continuous process if T_i is totally inaccessible and that V^i is of the form $V^i = E(f_i | F_{T_i^-}) \cdot \chi_{[T_i, \infty[}$ if T_i is predictable (see 8.10 Corollary 4).

Consequently we get for the k'th dyadic partition $t(k,0), \ldots, t(k,n)$ of the interval $[0,t]$:

$$\lim_k \Sigma_j (M_{t(k,j+1)} - M_{t(k,j)})^2 =$$
$$= \lim_k \Sigma_j [\Sigma_i (M^i_{t(k,j+1)} - M^i_{t(k,j)})]^2$$
$$= \lim_k \Sigma_j \Sigma_i (M^i_{t(k,j+1)} - M^i_{t(k,j)})^2$$
$$= \Sigma_{s \leq t} (\Delta M_s)^2 \ .$$

Taking into account that $S_k = [\Sigma_j (M_{t(k,j+1)} - M_{t(k,j)})^2]^{1/2}$ is smaller than the integrable function $2\Sigma_i (|f_i| + |V^i|_\infty)$ for every k, we have $\lim_k S_k = [\Sigma_{s \leq t} (\Delta M_s)^2]^{1/2}$ in the $L^1(P)$-norm by virtue of the Dominated Convergence Theorem. In view of 12.12 Corollary and 4.18 Corollary 2, this completes the proof.

13.7 Definition: The subspace cSM of SM is defined as the space of all continuous summable martingales, zero at time $t = 0$.

Recall that the subspace dSM of SM of purely discontinuous summable martingales was defined as the closure of H_d^2 in SM.

13.8 Theorem: The Banach space SM is the (topological) direct sum of its subspaces cSM and dSM.

Proof: We first show that there is a continuous projection Π of SM onto dSM. Let $M = (M_t)$ be an element of SM and (S_n) be a sequence of stopping times, exhausting the jumps of M. There is a sequence (T_n) of totally inaccessible or predictable stopping times with disjoint graphs, such that the union of the graphs of the sequence (T_n) contains the union of the graphs of the sequence (S_n). Define the martingale M^n as

$$M^n = \Delta M_{T_n} \chi_{[T_n, \infty[} - V^n \quad n = 1, 2, \ldots, \quad \text{where} \quad \Delta M_{T_n} = M_{T_n} - M_{T_n-} \quad (M_{0-} = 0)$$

and V^n is the compensating process $M_{T_n} \chi_{[T_n, \infty[}$. Define M as the sum $M = \Sigma_n M^n$. M is well defined, at least for square integrable martingales M because Π, restricted to the Hilbert space of square integrable martingales, is the orthogonal projection on the closed subspace of purely discontinuous square integrable martingales (see 11.8 Corollary 1). Since H_d^2 is dense in dSM, the assertion that Π is a continuous projection onto dSM is easily deduced from Proposition 13.6 and 13.6 Corollary.

It is clear that cSM is a closed subspace of SM, and using an optional stopping argument, Proposition 13.2 shows that the space of continuous square integrable martingales is dense in cSM. According to 11.7 and 11.8 Corollary 1, $Id - \Pi$ restricted to the Hilbert space of square integrable martingales is the orthogonal projection onto the subspace of continuous square integrable martingales, hence $Id - \Pi$ is a continuous projection onto cSM.

This completes the proof of the theorem.

In [20] the dual BMO of H^1 is completely characterized as the space of all $g \in L^2(P)$ such that the right continuous modification $g = (G_t)$ of $t \to E(g|F_t)$ satisfies the following condition: $E[(G_\infty - G_{T-})^2|F_T]$ is (P almost surely) bounded by a constant c, uniformly for all stopping times T. For the sake of completeness we prove the following:

13.9 Proposition: The dual of H^1 can be identified with a subspace BMO of $L^2(P)$, satisfying:

i) If f is an element of H^1 and g an element of BMO and if we denote by $F = (F_t)$ resp. $G = (G_t)$ the right continuous modifications of $t \to E(f|F_t)$ resp. $t \to E(g|F_t)$, then the process $FG = (F_t G_t)$ is a local quasimartingale.

ii) For every $g \in BMO$ there is a constant K, such that $|\Delta G_T| = |G_T - G_{T-}|$ $(G_{0-} = 0)$ is bounded by K uniformly for all stopping times T.

iii) Denoting by $<F,G> = (<F,G>_t)$ the unique predictable process of bounded variation, such that $FG - <F,G>$ is a local martingale, zero at time $t = 0$, $<F,G>$ is of integrable variation for every $f \in H^1$ and $g \in BMO$, and the dual bilinear form on the pair

(H^1, BMO) is given by $<f,g> = E(<F,G>_\infty)$.

Proof: Let ρ be a continuous linear form on H^1 of norm ≤ 1. According to 13.2 Corollary $L^2(P)$ is a dense subspace of H^1 and the embedding of $L^2(P)$ into H^1 is continuous. Consequently there is a unique element $g \in L^2(P)$ such that $\rho(f) = E(fg)$ holds for all elements $f \in L^2(P)$, and ρ is the continuous linear extension of $f \to E(fg)$ to H^1. This observation allows us to identify the dual of H^1 with a subspace BMO of $L^2(P)$. Note that BMO contains the space $L^\infty(P)$.

Let now g be an element of BMO of norm ≤ 1 and denote by $G = (G_t)$ the right continuous modification of $t \to E(g|F_t)$. Let us first prove that G satisfies ii). Let T be a stopping time and A an element of F_T. Denote by M the martingale $M = \chi_A \chi_{[T,\infty[} - V$, where V is the compensating process of $\chi_A \chi_{[T,\infty[}$. M is a square integrable martingale and at the same time a well measurable process of integrable variation with an SV-norm smaller than $2 P(A)$. Since the SM-norm is smaller than the SV-norm, there is a constant K (which is independent of the particular stopping time T), such that the H^1-norm $p_1(M_\infty)$ of M_∞ is smaller than $K P(A)$ (12.12 Corollary). Taking into account 11.8 Corollary 2, this yields
$$K P(A) \geq |E(M_\infty g)| = |E(\Delta M_T \Delta G_T)|.$$

If T is totally inaccessible, $\Delta M_T = \chi_A$ and if T is predictable $\Delta M_T = \chi_A - E(\chi_A|F_{T-})$ and $\Delta G_T = E(g|F_t) - E(g|F_{T-})$ hold, i.e. in both cases we have $E(\Delta M_T \Delta G_T) = E(\chi_A \Delta G_T)$.

Thus we have $|E(\chi_A \Delta G_T)| \leq K P(A)$ for every $A \in F_T$ if T is totally inaccessible or predictable. Consequently $|\Delta G_T| \leq K$ holds in these cases, which yields by virtue of Theorem 6.5 $|\Delta G_T| \leq K$ uniformly for every stopping time T. This completes the proof of assertion ii).

The assertion i) now follows easily from the fact that for an element $f \in H^1$ and $g \in L^\infty(P)$ the process $(F_t G_t)$ is a quasimartingale (see 12.3 e)), and that for an element $g \in BMO$ the martingale $G = (G_t)$ is locally bounded in modulus by virtue of ii).

Furthermore for elements $g \in L^\infty(P)$ we have $E(fg) = E(<F,G>_\infty)$ (see 12.7 d)), the assertion iii) follows from the facts that an element $g \in BMO$ represents a continuous linear form on H^1 and that $G = (G_t)$ is locally bounded according to ii), together with Theorem 13.2.

References

1. Blumenthal R.M. and Getoor R.K., Markov Processes and Potential Theory, (Academic Press, New York and London 1968).

2. Burkholder D.L. Martingale transforms, Ann. Math. Statist. 37 (1966) 1494-1504.

3. Davis B. On the integrability of the martingale square function, Israel J. Math. 8 (1970) 187-190.

4. Dellacherie C. Capacites et processus stochastiques, Ergebn. der Math. vol 67 (Springer, Berlin 1972).

5. Doleans-Dade C. and Meyer P.A., Integrales stochastiques par rapport aux martingales locales, in Seminaire de Prob. IV Lecture Notes in Mathematics 124 (Springer Berlin 1970).

6. Doob J.L. Stochastic processes (Wiley, New York, 1953).

7. Dunford N. and Schwartz J.T., Linear Operators Part I: General Theory, (Wiley, New York 1958).

8. Fefferman Ch. Characterisations of bounded mean oscillation, Bull. Am. Math. Soc. 77 (1971) 587-588.

9. Fisk D.L. Quasi-martingales, Trans. Am. Math. Soc. 120 (1965) 369-389.

10. Follmer H. The exit measure of a supermartingale, Z. Wahrscheinlichkeitstheorie Verw. Gebiete 21 (1972) 154-166.

11. Garsia A.M. The Burgess Davis inequalities via Fefferman's inequality, Ark. Math. 11 (1973) 229-237.

12. Herz C.S. Bounded mean oscillation and regulated martingales, T.A.M.S. 192 (6) (1974) 199-215.

13. Ito K. Stochastic integral, Proc. Imp. Acad. Tokyo 20, (1944) 519-524.

14. Kacmarc S. and Steinhaus H., Theorie der Orthogonalreihen (Warzawa-Lwow Monografje Matematyozne 1935).

15   Kunita H. and Watanabe S., On square integrable martingales, Nagoya Math. J. 30 (1967) 209-245.

16   McKean jr. H.P.   Stochastic integrals (Academic Press, New York and London, 1969).

17   Metivier M. and Pellaumail J., On Doleans-Follmer's measure for Quasi-Martingales, Ill. J. Math. 77 (1975) 491-504.

18   Meyer P.A.   Martingales and stochastic integrals I, Lecture Notes in Mathematics 284 (Springer, Berlin 1972).

19   Meyer P.A.   Probability and Potentials, (Blaisdell, Waltham Mass. 1966).

20   Meyer P.A.   Un course sur les integrales stochastiques, in Seminaire de Prob. X, Lecture Notes in Mathematics 511 (Springer, Berlin 1976) 245-400.

21   Neveu J.   Discrete-Parameter Martingales (North Holland 1975).

22   Orey S.   F-processes, in Proc. 5th Berkeley Symp. on Mathematical Statistics and Probability II, vol. 1, (Univ. of California Press, Berkeley, Cal. 1965) 301-314.

23   Pellaumail J.   Une nouvelle construction de l'integrale stochastique. Percolation et supraconductivite. These, (Universite de Rennes 1972).

24   Rao K.M.   Quasimartingales, Math. Scand. 24 (1969) 79-92.

25   Schaefer H.H.   Banach lattices and positive operators. Grundlehren der math. Wissenschaften in Einzeldarst. Band 215 (Springer, Berlin 1974).

26   Schaefer H.H.   Topological vector spaces (Macmillan, New York (1966).

27   Wiener N.   Differential space, J. Math. Phys. Math. Inst. Tech. 2, (1923) 131-174.

28   Yen K.A. and Yoerp Ch., Representation des martingales comme integrales stochastiques des processus optionels, in Seminaire de Prob. X, Lecture Notes in Mathematics 511 (Springer, Berlin 1976) 422-480.

29   Yor M.   Representation integrale des martingales de carre integrable, C.R. Acad. Sc. Paris t. 282 (1976) 899-901.

30 Zygmund A. Trigonometric Series, vol I and II, (Cambridge University Press, 1959).

Index of symbols

§1

Ω, 1

Σ, 1

R, R_+, 1

P, $L^p(P)$, 1

$E(f|\Sigma_0)$, 1

$\|f\|_p$, $\|f\|_\infty$, 1

$X = (X_t)$, 2

$X(\omega)$, 2

$B \otimes F$, 2

$M(F)$, 3

§2

$V = (V_t)$, $|V| = (|V|_t)$, 5

$V(Z) = (V(Z)_t)$, 5

$\int_{[o,t]} Z_s dV_s$, 5

$|V|_\infty$, $V(Z)_\infty$, 6

SV, 6

$\Sigma = B \otimes F$, 6

μ_V, $|\mu_V|$, 6

$sca(\Sigma)$, 7

I_V, $I_V(Z)$, 9

§3

F_{t+}, F_{t-}, 11

$(\Omega, F, P(F_t), T)$, 12

Σ_0, 12

$\{S \leq t\}$, 13

F_S, 13

X_S, X^S, 14

§4

N, 18

$X(Z) = (X(Z)_n)$, 18

U_a^b, 20

M, M^p, 29

H^1, 30

E^k, E_m^k, 30

$\bar{p}_1$, $\bar{p}_2$, p_1, p_2, 31

q_1, q_2, $\bar{q}_2$, 31

§6

F_{T-} , 44

$]S,T]$, $[S,T[$, $[S]$, 47

T_A , t_A , 47

§7

$D(A)$, 53

Σ_a , Σ_p , Σ_w , 54

$B^\infty(\Sigma)$, $B^\infty(\Sigma_p)$, $B^\infty(\Sigma_w)$, $B^\infty(\Sigma_a)$, 65

π_w , π_a , π_p , 66

§8

$\mathrm{sca}(\Sigma_w)$, $\mathrm{sca}(\Sigma_a)$, $\mathrm{sca}(\Sigma_p)$, 69

π'_x ($x = w,a,p$) , 69, 75

wSV, aSV, pSV, 75

§9

R, 86

μ_X , 87

$\mathrm{ba}(R)$, 89

$\mathrm{sba}(R)$, 89

§10

A, Σ , 99

$\mathrm{ba}(A)$, $\mathrm{ba}(\Sigma)$, 99

$\mathrm{ca}(A)$, $\mathrm{ba}(\Sigma)$, 99

M, vM, svM, 100

$S(A)$, $S(\Sigma)$, 102

$B(A)$, $B(\Sigma)$, 103

N , 103

$B^\infty(M)$, 103

$\langle M, x' \rangle$, 103, 101

$L^1(M)$, 105

§11

A, 112

I_X , 112

$\langle I_X, f \rangle$, 115

I'_X , 115

$\langle X, F \rangle = (\langle X, F \rangle_t)$, 115

$X(Z) = (X(Z)_t)$, 116

H^2_c , H^2_d , H^2_T , 120

§12

$\langle X, F \rangle = (\langle X, F \rangle)_t$, 129

SP, SM , 134

M^* , 138

H^1 , 139

§13

BMO, 146

cSM, dSM, 151

Subject index

Accessible set, 54

Condition (S), 88

Convergence in M-measure, 101

Compensating process, 120

Debut, 53

Doob-

-inequalities, 23, 24

-optional sampling theorem, 28

-upcrossing lemma, 20

Doob-Meyer decomposition, 91

E-process, 3

   modification of a -- , 3

Evanescent set, 3

Fefferman's inequality, 33, 37

First hitting time, 54

Indistinguishable, 2

Integrable, 102

- w.r. to a locally summable

- process, 130

- locally -- , 130

Integral, 101

Jump of a stochastic process, 63

Lattice ideal, 89

Lattice orthogonal, 89

Martingale

   - transform, 18

   local - , 95

   p-integrable - , 16

   purely discontinuous - , 120

   sub - , 16

   super - , 16

   uniformly integrable - , 16

Measure, 1

   - of finite variation, 100

   E-valued - , 1

   probability - , 1

   semivariation of a - , 100

   variation of a - , 100

Null set, 101

Potential, 91

Predictable set, 54

   finite - , 132

Probability space, 1

Process of integrable variation, 6

   real measure generated by a - , 36

Progressive set, 12

Projection

   accessible - , 66

   dual Σ_x-measurable - $(x=w,a,p)$, 75

   predictable - , 66

   well measurable - , 66

Quasimartingale, 80

   local - , 126

   measure associated with a - , 87

Quasipotential, 91

Reverse supermartingale, 26

Riesz decomposition, 91

Semimartingale, 126

Sequentially compact, 100

Stable under stopping, 119

Stochastic base, 12

Stochastic integral, 5

- w.r. to a locally summable process, 132

- w.r. to a p-summable process, 115

Stochastic interval, 47

   bounded - , 86

Stochastic measure, 7

   - generated by a p-summable process, 113

Stochastic process, 2

- locally of class (D), ,95

- of bounded variation, 5

- of class (D), 92

- representable as stochastic integral, 144

- stopped at S, 14

   adapted - , 12

   base of a - , 2

   constant - , 2

   continuous - , 2

   evanescent - , 2

   increasing - , 5

   measurable - , 2

   locally summable - , 125

   path of a - , 2

   p-summable - , 113

   real - , 2

   right, left continuous - , 2

   state space of a - , 2

   summable - , 113

Stopping time, 13

- charged by a stochastic process, 63

   accessible - , 47

   graph of a - , 47

   predictable - , 46

   reducing - , 126

restriction of a - , 47

sequence of -s exhausting the jumps of a process, 63

totally inaccessible - , 47

Weak topology on ca(Σ) , 100

Well measurable set , 54